Strategies for Business and Technical Writing

Second Edition

Strategies for Business and Technical Writing

Second Edition

Kevin J. Harty
La Salle University

Harcourt Brace Jovanovich, Publishers

San Diego   New York   Chicago   Atlanta   Washington, D.C.
London   Sydney   Toronto

ISBN: 0-15-583925-X

Library of Congress Catalog Card Number: 84-81518

Printed in the United States of America

For Jamie
Docendo discimus.

Preface

*T*his Second Edition of *Strategies for Business and Technical Writing* is a thorough reworking of the first. Like its predecessor, it can be used in three ways—as a supplement to a business or technical writing text; as a supplement to a composition text or handbook; or on its own as a text or reference book. The 14 new selections broaden the book's coverage of business and technical writing and increase its usefulness as a text and reference. In particular, the new selections

- make explicit the first edition's emphasis on the *process* of writing
- offer new perspectives on such topics as audience analysis and adaptation, graphics, letters of application, and resumes
- extend coverage to such important new topics as nonsexist writing, the Plain English revolution, and proposal writing.

Strategies for Business and Technical Writing is designed to appeal to practical-minded instructors, students, and business people. Its contributors write with the kind of purpose and understanding that come only from a career in business and technical communications. As a result, the selections in this anthology not only teach professional writing; they are themselves fine examples of the genre.

All courses in business and technical writing should firmly implant one idea in students' minds: *All* successful writing consists of clear and effective prose. Students should recognize that business and technical writing differs in important ways from the expository prose usually taught in freshman composition courses. Since business and technical writers communicate with multiple audiences and with a variety of intentions, learning to write well for business and industry requires some adjustments in technique. Therefore, in addition to covering such broad problems as style, jargon, and diction, *Strategies* includes comprehensive discussions of specific forms—among others, resumes, application letters, letters that sell, memoranda, formal reports, and proposals. Students learn the specific techniques used by successful business and industry executives who communicate information as part of their jobs. In Part 1, for example, Herbert Popper sets forth six guidelines for professional writers that his long experience has distilled from many possibilities. And in Part 3, Marvin Swift recreates the time-saving process by which an efficient manager composes a routine memorandum. In addition to the essays, readers of *Strategies* will find the bibliography in Part 6 helpful in locating more information on specific topics for reference or for library research.

The essays in *Strategies* have survived the most rigorous scrutiny—that of my students in seminars and courses in business and technical writing at La

Salle University, Temple University, Rhode Island College, the Federal Reserve Bank of Philadelphia, Philadelphia Life Insurance Company, First Pennsylvania Bank, Blue Cross of Greater Philadelphia, and Gino's Inc. I am grateful to these former students for being such willing and helpful critics.

I would also like to thank several others for their advice and assistance while this revision was in the making. Vicki Goldsmith of Northern Michigan University, Kathryn Harris of Arizona State University, and Thomas Willard of the University of Arizona wrote detailed reviews of the manuscript for this edition from which I have benefited greatly. At La Salle University, Margot Soven offered valuable advice on the contents of this edition; John Keenan, department chairman, has been mentor, colleague, and friend—to him I owe fourfold thanks; John Christopher Kleis has been an advocate for this and other projects; and Margaret Grzesiak, the former coordinator of interlibrary loans at La Salle, was tireless in her efforts to meet all my requests.

The first edition of *Strategies* owed much to the efforts of two editors at Harcourt Brace Jovanovich, Drake Bush and Bob Beitcher. Their guidance informs this edition, together with that of my current editor, Paul Nockleby, and his associates at HBJ. Finally, the dedication of this revision records with thanks my greatest debt.

K. J. H.

Contents

A worthwhile rule for all business and technical communication: Consider the customer or reader first, and put yourself in
his or her place when attempting to follow up on problems
and complaints.

Writers should (1) know their audience, (2) mobilize their subconscious, (3) exploit their momentum, (4) avoid verbosity, (5)
use examples, and (6) know themselves and their writing.

Writers need to keep purpose, audience, format, evidence, and
organization in mind.

The work schedule for any writing task ought to be 15% worrying, 10% planning, 25% writing, 45% revising, and 5% proofreading.

Marvin H. Swift
Clear writing means clear thinking means . . .   **145**

> By analyzing how a manager reworks and rethinks a routine
> memorandum, Swift addresses a constant challenge to man-
> agement—the clear and accurate expression of a well-focused
> message.

4

Reports and proposals   **151**

J. C. Mathes and Dwight W. Stevenson
Audience analysis: the problem and a solution   **152**

> Identification and analysis of the backgrounds and needs of
> primary and secondary audiences.

Richard W. Dodge
What to report   **169**

> The results of a widely cited study of the reading habits of
> Westinghouse managers: They prefer abstracts to full-length
> reports.

Christian K. Arnold
The writing of abstracts   **177**

> Key elements of effective abstracts.

Charles T. Brusaw, Gerald J. Alred, and Walter E. Oliu
Graphs, illustrations, and tables   **182**

> Easy-to-follow guidelines for producing three of the most
> common visual aids to printed documents.

Darrell Huff
How to lie with statistics   **198**

> A classic article about the way statistics can be misused to
> make the insignificant seem important.

6

Annotated bibliography

More than 275 additional sources of information.

Strategies for Business and Technical Writing

Second Edition

Introduction

Professional and technical occupations employ over half the college-trained people in the United States, and persons in those occupations wrote on the average 29% of total work time. . . . Everyone in a technical or professional occupation who is counted in the present survey wrote on the job . . . [and] two-thirds of our sample of people in technical and professional occupations wrote at least one full working day out of every five.[1]

Ability to express ideas cogently and goals persuasively—in plain English—is the most important skill to leadership. I know of no greater obstacle to the progress of good ideas and good people than the inability to compose a plain English sentence.[2]

Whether you are planning a career in business, industry, or some technical field, or whether you already have such a career, your training may leave you ill-prepared for what may perhaps be the most difficult part of your job. You may be an excellent accountant, engineer, or scientist, but all your experience may be of little help when you sit down to write. If so, you are not alone in being frustrated by the writing demands of your job. The fact is that writing is essential to most technical and professional occupations, and one measure for advancement within such occupations is the ability to communicate effectively both in person and on paper. Good writing can mean the difference between winning or losing a job interview, a big sale, a pleased customer, or a challenging new job (with a higher salary).

The selections in this book are devoted to ways in which you can make your professional letters, reports, and memoranda more effective, and they represent some of the best advice veterans of business and technical writing have to offer. None of the selections, however, offers a quick cure for writing ills. There is no such cure. Like any worthwhile skill, good professional writing requires time, practice, and, most of all, discipline.

To begin with, there are only two kinds of professional writing:

---

[1] Lester Faigley and Thomas P. Miller, "What We Learn from Writing on the Job," *College English,* 44 (October, 1982), 560.

[2] John D. deButts, former Chairman, AT&T. Quoted as the epigraph to the third edition of the Bell Laboratories' *Editorial Style Guide* (Whippany, NJ, 1979).

- clear, effective writing that meets the combined needs of the reader and the writer

- bad writing.

Bad writing is unclear and ineffective; it wastes time and money. Even worse, it ignores its readers, and it usually raises more questions than it answers.

To be effective writers, each of us needs to develop a variety of strategies for the different writing situations we face. We must also respect writing as something more than words typed or written on the page or fed into a dicta-phone or word processor. Writing takes thought and planning.

Three key questions

Different writing situations present us with different challenges. One of the ways of meeting these challenges is to begin the writing process by asking ourselves these three key questions before we actually start to write:

- Exactly who is my audience?

- What is the most important thing I want to tell my audience?

- What is the best way of making sure my audience understands what I have to say?

Exactly who is my audience? Whenever we write, some person or per-sons—across the hall, the city, the country, or even the world—will read what we have written an hour, a day, or maybe even a year later. Most likely, our reader or readers will read our letter or report in their own offices or homes. That is, our work will have to speak for itself, since we won't be on hand to explain anything that is unclear or that doesn't make sense.

Whenever we sit down to write, we are the expert. That is in large part why we are the writer and not the reader. Our readers, in turn, depend on us for a clear, effectively written message. Although these facts seem obvious, business, industrial, and technical practice shows that the single most common cause of bad writing is ignoring the audience. There are any number of ways to ignore our readers. We can talk over their heads. We can talk down to them. We can beat around the bush. We can leave out important details they need to know. We can do any or all of these things—and, as a result, we can under-mine our credibility with our readers.

That is why, when you next sit down to write, it is important to ask yourself what you know about your audience—about your audience's situ-ation, about your audience's biases, about your audience's ignorance or built-in resistance, about what it is you hope to accomplish with this piece of writ-ing, about your audience's potential reaction to this particular letter, report, or memorandum. To paraphrase the golden rule, when you sit down to write, make sure you are prepared to write unto others the way you would have them write unto you.

What is the most important thing I want to tell my audience? Whether you write one page or one hundred pages, you should be able to tell your reader in about 25 words what your most important idea is. If you can't, you aren't yet ready to write, since you haven't figured out what you are trying to say. In 25 words you may not be able to explain every reason for a decision or a proposal, but 25 words should be enough for the decision or proposal itself. For instance, you may have several reasons for recommending that your company replace its Brand X word processors with Brand Y. Nevertheless, you can make that recommendation in 22 words flat:

> I recommend that we immediately replace all our Brand X word processors with new Brand Y units for the reasons presented below.

In short, you can cut through all the fluff and folderol—all the bull—and come right out with your recommendation. Then, in the body of your memorandum or report, you can present the relevant facts. But get your main point out first in clear, unadorned prose so that your reader knows from the start what you are talking about. Busy people dislike having their time wasted, and they hate surprises. For these reasons, in writing, the bottom line should also be the top line.

What is the best way of making sure my audience understands what I have to say? Different situations require different strategies. After answering the first two questions, you may decide that you shouldn't write at all—that you should phone instead or make a personal visit. But once you do commit yourself to writing, you then need to determine how to put this particular letter, memorandum, or report together. You will need to use one strategy for good news and a different one for bad news, one for giving information and another for changing someone's mind, one for accepting a job offer and another for turning the offer down. But you will need in each case to get your main idea out in as few words as possible and then to defend or elaborate on that idea as clearly and effectively as possible.

To illustrate the importance of answering these three key questions properly, let's look at a brief excerpt from a booklet formerly used by the New York State Income Tax Bureau:

> The return for the period before the change of residence must include all items of income, gain, loss or deduction accrued to the taxpayer up to the time of his change of residence. This includes any amounts not otherwise includable in the return because of an election to report income on an installment basis.
>
> Stated another way, the return for the period prior to the change of residence must be made on the accrual basis whether or not that is the taxpayer's established method of reporting. However, in the case of a taxpayer changing from nonresident to resident status, these accruals

need not be made with respect to items derived from or connected with New York sources.

How do you react to this information? If you were a taxpayer in need of quick and concise information to solve a particular problem, you would surely be disappointed and even annoyed. Tax forms need not be complicated. Since ordinary people normally fill out such forms, they should be written for the average taxpayer, not for a trained accountant.

Furthermore, these two paragraphs did not fulfill their purpose: to tell taxpayers how to adjust their returns when changing residence. Although the grammar is correct, the message is certainly not effective. Among other faults, the writer forgot to ask, "Who is my audience?" Unfamiliar words or usages such as *includable, accrual basis,* and *election* appear without explanation. If taxpayers cannot understand such language, they will complete their returns incorrectly and cheat either themselves or the state out of money. So, in the end, these two paragraphs—and others like them in the tax booklet—will have missed their mark, and unnecessary complications will be the result.

Guidelines for effective writing

To make sure you answer the three key questions correctly, keep these five guidelines in mind:

• *Put your main idea up front.* Then give whatever background or reasons your reader needs to understand your main idea.

• *Avoid using technical terms with nontechnical readers.* If you are an accountant writing to another accountant, use all the technical terminology you need. But if you are an accountant writing to a lay person, use ordinary language that the lay person will understand. If you must use technical terms, define and explain them carefully.

• *Limit the length and complexity of your sentences.* Whenever possible, try to include only one main idea in each sentence. More mechanically, try also to limit your sentences to 25 words (or about 2½ typed lines). Generally speaking, the shorter the sentences, the fewer the difficulties you and your reader will get into. To ensure that your sentences are easily understood, move the main subject and verb to the front of the sentence and cut out any unnecessary words.

• *Limit the length and content of your paragraphs.* Nothing intimidates a reader more than a half-page-long block of dense prose. Try keeping your paragraphs down to four or five sentences. If you really need a longer paragraph, use indentation or enumeration so that your subpoints stand out from the pack.

• *Give your reader all necessary information in a clear and unmistakable way.* If your reader needs to let you know something by June 12, tell the reader so plainly. If you want your reader to check back with you for additional information, make sure you provide your address or telephone number—and your travel or vacation schedule if you are going to be out of the office. If you are attaching or enclosing anything, specify exactly what. If your letter is a follow-up to some previous correspondence or to a recurring problem, make sure the reader knows this too.

The 28 selections that follow suggest many ways in which you can make your professional writing more effective. They also suggest ways you can make the process more enjoyable. If you view your professional writing as a chore, the effectiveness of your writing will be limited. But the more you get into the habit of asking the three key questions before you write your memoranda, letters, and reports, the easier, the more effective—and the more enjoyable—writing will become for you.

1

Process as well as product

When we tell customers, "We hear you," we'd better be listening

JOHN D. deBUTTS

When he wrote this article, John D. deButts was
Chairman of the Board of the American
Telephone and Telegraph Company (AT&T); he
has since retired.

*I*f you were to sit at my desk and read
and answer the mail I get about telephone service, you'd come to realize, if you
don't already, how much "little things" mean; how sometimes lifelong impres-
sions of the character of our business are formed on the basis of a single
contact with just one employee. Rather significant, I'd say, when you consider
that the Bell System employs more than a million men and women.

You know, it's only natural when a customer feels service is poor or when
he thinks his bill is too high that he is going to register a complaint with
someone. But in all my years in this business, I can't recall ever receiving a
letter that complained solely about cost or service. No, the key element in all
the letters of complaint I get is how the customer was treated by an employee
of the company. As a matter of fact, I'm sometimes amazed at just how toler-
ant our customers are of situations that to me would seem intolerable, before
poor treatment by an employee finally pushes them to write to me or to one of
the other officers, or to a public utility commissioner.

Reprinted from *Bell Telephone Magazine*, 52 (September-October, 1973), pp. 2–5, by permission
of the American Telephone and Telegraph Company.

So that I don't give the wrong impression, let me add that intermixed with those complaints are a goodly number of commendation letters. And commendation letters, of course, always talk about how an employee treated the customer in a manner so unusually satisfying that the customer felt compelled to write to me about it.

I'd like to share portions of some of those letters with you. Here's one from a man in New Jersey:

> Today, Saturday, at approximately 9:30 a.m., my next door neighbor rang my door bell to advise me that either my power or telephone line had pulled loose from my home and was lying across the street where cars coming and going could have caused a serious problem.
>
> I went out to investigate, when, at the same time, one of your New Jersey Telephone trucks was approaching. He immediately pulled his truck to the curb, hopped out and raised the line from the road. Without any words from me, he took his ladder down from the truck and proceeded to reinstall the line to the house and make necessary repairs.
>
> I then spoke to him for the first time. He had been in the vicinity making a phone installation, and spotted the trouble . . . I thanked him . . . I feel that AT&T and New Jersey Bell should be mighty proud to have a man like that working for you.

The customer was right. We *are* proud to have such a man—the kind of employee who epitomizes the Spirit of Service we used to talk so much about in the old days, and who possesses that "sense of proprietorship" we've been talking about more recently. We wish all our employees showed the same sense of judgment and concern for a customer's problem. However, in an organization as large as ours, there are bound to be some weak links.

Here's a letter from a customer who also happens to be a share owner:

> My telephone has been out of order for the past four days. On April 8, 1973 at 7:00 p.m., I made my fifth call to your service department and was told, as I had been told four times previously, 'We'll take care of it.' I asked for his name and he refused to give it. I asked for the identity of his supervisor and again was refused . . . I asked, 'Do I have to go to New York to get my phone service restored?' He retorted, 'You can call whoever the hell you want, but it won't get your service a damn minute sooner.' Further discussion was pointless and I hung up.
>
> The policy of permitting employees that have contact with customers to retain anonymity from behind Ma Bell's skirts is poorly conceived and poor customer relations. I cannot imagine any other company operating with this policy.

Now, consider this letter for a moment. The customer had been without service going on five days when he wrote. Subsequent investigation showed that one of our own cable crews working in the area was causing the problem, and that the customer was without service for still another day. Repair had been diligently searching for the cause of the problem from the time of the customer's first call, but no one thought to tell the customer what was being done. Obviously, the customer has every right to complain about being without service, but what is it that he is really complaining about here? Not lack of service, but an apparent lack of concern for his predicament, and the rude treatment he received from an employee.

When you get right down to it, there's really nothing very complicated about maintaining good customer relations. The guideline has been around a long time. Just follow the Golden Rule. If we treat the customer as we ourselves would like to be treated under the same set of circumstances, our actions are bound to meet with his approval. When I look through the letters I get from customers, I simply ask myself, "How would I like to be treated if I were in that customer's shoes?" The answers are usually fairly obvious.

Back when I was president of Illinois Bell, we had a saying there that "You not only have to give good service, but you've got to convince the customer that he's getting it." Here's a letter to illustrate what I mean.

> I am sorry to bother you but you should know about the poor service I had on the weekend.
> Saturday, April 28, about noontime, I found I could not use my phone . . . I reported on my neighbor's phone, immediately. I asked them to please fix it right away as I am an elderly lady—83 years old—and live alone. They said they would come before 5 p.m. I waited for them all Saturday and at 6 p.m. a man came and said he would come back later. He never came into my home. I waited for him until 11 p.m. and then went to bed. This was very upsetting and frightening for me and I didn't sleep. After I called again, they finally came . . . at 3 p.m., Sunday. I have large telephone bills and want better service than this.

In this case, Repair dispatched a man to investigate, and he found a short circuit in a cable on a pole outside the home. He worked on the trouble from Saturday afternoon into the early evening but couldn't finish the job and told the lady he'd return. He came back the next day, completed the splicing by early afternoon and left. The lady, not realizing that the repair had been completed, called again from the neighbor's phone. Another man was sent to check the problem, found everything in order, and reported to the lady that her phone was back in service. At which point, the lady took up her pen to write and tell us how we had let her down.

Now, I'm entirely sympathetic with this lady and can understand how such a thing can occur. A colleague of mine had a similar experience with an air-conditioning repairman. The man came into the home, looked briefly at the air conditioner, put the cover back on, picked up his tools and left. My friend's wife thought he had gone to get something from his truck, but when she looked out, the truck was gone. She waited all day for him to return, but he never came back. A second call to the repair shop the next day brought the information that a new part had to be ordered and, until it was received, the repair couldn't be completed. How easy it would have been, in both of these cases, for the repairman to give the customer a progress report.

Here's an excerpt from a letter I received from another share owner, one who had recently moved, relating an experience he had in merely inquiring about the types and cost of phone service:

> . . . I encountered a faceless voice that first demanded answers to a long list of questions, many of them personal, very few of them that could have any bearing at all on the answer to my inquiry. Rightly or wrongly, I became irritated by the manner of my interrogator and when I recognized she was more interested in obtaining answers than she was in being helpful, and the questions had become so absurdly irrelevant to my request . . . I terminated the interview.

Now service representatives, obviously, must ask some questions if they are to do their jobs properly. But they should be flexible enough to know when to draw the line. When a customer begins to show irritation, as this one did, it's time to put aside the set routine. Fewer customers would have to be put through such prolonged fact-gathering if we spent just a little more time listening to what it is that the customer wants from us.

Finally, in the interest of balance, here's a note with a happy ending— from a customer who had a disappointing experience with one operator, only to have her faith restored later by the patient, resourceful service she received from another operator.

In contrast to her experience with the first operator, the customer had this to say about the second:

> . . . [She] listened to me, investigated the situation and had the courtesy to call me and explain that the number was out of order, that no report had been made to that effect earlier, but that she had done so.

Note the key word in this letter. The operator *listened* to her. The operator not only heard a problem described, but recognized how important it was for this customer to determine why she could not complete this particular call. The second operator, like the first, could have let the matter drop, but didn't. She followed through, determined the phone was out of order and reported it to the appropriate people.

Getting away from the letters for a minute, far more of our customer contacts come by phone. And of course, a certain number are in person. Our Operations-Commercial group ran a check on some 200 business offices around the country recently to see how "reachable" the business office manager was to the customer.

Well, I'm happy to say that the calls placed were handled properly more than 80 per cent of the time. If the caller was not passed along, he was not only told why, but also informed when he could expect the manager to return the call. Although I'm happy that the majority of the calls did go through, I'm equally concerned about the 20 per cent that didn't, and I worry about why they didn't.

We've got to stop building walls around our managers. These are the people who have the authority to correct the things the customers are complaining about. In all my years in this business, I've always had my home telephone number published in the directory. I know this isn't generally true of executives in other industries, but I've always felt it was the right thing to do. Over the years, I've received my share of crank calls, but I've also gotten calls from people just seeking help with their phone service. I've always been more than happy to help solve their problems whenever I can.

The Pacific Company, one time, made a study of some 150 written customer complaints. These complaints had been mailed either to company executives at Pacific, or directly to a public utility commissioner. In tracing the complaints back to see what kind of treatment the customer got along the way, the company discovered that more than half of the letter writers, 58 per cent, as I recall, had contact with only one employee before taking up their pens.

Like the 20 per cent of the calls that didn't get through to the managers in the Operations test, these letter writers didn't get the chance to have their problems reviewed by supervisors or managers at the local level. Well, you can bet that, following the written complaint, the customer got the attention from the people who should have given it to him the first time.

Again, don't misunderstand. I'm not suggesting that we capitulate to unreasonable demands, simply to curtail the number of complaints this company receives. We have rules and regulations to live by, and generally they must be observed. What does concern me, though, is that too often our employees hide behind the rule book and fail to honor some very sound requests simply because they are afraid to risk doing what they know is right, because it's not covered by the rules. In the final analysis, rules are there to guide us, not to govern us.

Take the case of the man in Brainerd, Minnesota, one of the very few places left in the country where equipment shortages still require 8-party service. As he was a year-round resident in an area primarily inhabited by summer vacationists, he asked if it wouldn't be possible to tie his line in with some of theirs. He reasoned that for most of the year, at least, when they were gone, his telephone wouldn't be ringing as often. You can't fault the logic. But do you know what our initial response was—absolutely not! You know why? If the other permanent residents found out, they'd want similar consideration. By the way, once he took the trouble to write me about his suggestion, we took another look at the rule book and found no reason why we shouldn't honor his request. He's a much happier man, today. And thinks a little better of us, too.

On the other hand, a Brooklyn customer, an elderly man, wrote telling how upset he was that his local bank had decided not to act as a collection agency for the company, and that he would now be forced to spend 70 cents in subway fares to come into the business office to pay his bill. While we couldn't force the bank to continue collecting his bill, as he suggested, we did ask the local business office manager to look into what could be done. Since the customer didn't believe in checking accounts, there was no safe way the manager could advise mailing in payments which would not incur costs almost as high as the subway fare to which the customer objected. In looking over the customer's payment record, the manager observed that over many years he always paid his bill promptly, and in full. In addition to providing this customer with the addresses of two other collection points nearer his home than the business office, the manager suggested, in light of the customer's fine credit record, that he remit payment every other month. Here's a manager willing to take a collection risk in an effort to do the right thing by this customer. And I find that very commendable indeed.

If we want to encourage employees to take such independent action in response to specific customer needs and requests, we, as managers, must set the climate. We'll never convince our customer contact-employees that we are sincerely interested in having them use their own good judgment in dealing with customers if we force a strict adherence to the rules and require double-checking of the most trivial matters up the line for higher level approval. While it is a manager's responsibility to see that his staff is adequately trained and can be relied upon to use common sense in dealing with customers, once that task is accomplished, managers should not keep intruding their presence. Following the careful selection, training and placement of the most competent employees in positions of trust and responsibility, nothing ruins initiative faster than to have managers who insist on second-guessing every decision and require that everything be run by the book. Again, we shouldn't throw the book away, but, I reiterate, the rules are there to guide us, not to govern us.

Over the past year, I've held more than 30 large, formal, employee meetings around the country and quite a number of smaller, informal, higher-level management meetings. At these meetings, I've stressed that I know of nothing more important to providing top-notch service than the *sense of proprietorship* each Bell System employee brings to his or her job.

Just recently, we embarked on a major national advertising campaign built around the simple phrase, "We Hear You." Through these three words we hope to convey to customers our determination to be responsive to their needs and concerns. Customer opinion, of course, will not be swayed one hair's breadth by any advertising campaign unless it is backed up by people who are ready and willing not only to listen, but to follow through with a satisfactory performance from the customer's point of view. When we say "We Hear You," it carries the promise that we will listen. Employees who have that sense of proprietorship will see to it that the promise is kept.

I remind you that where that sense of proprietorship has been lost, you and I have no more important responsibility than to restore it; and where it has not, no more important responsibility than to see that it is maintained.

When that's done, the customer will know that we're listening.

Six guidelines for fast, functional writing

HERBERT POPPER

Herbert Popper was a senior editor of the journal *Chemical Engineering*.

*S*ome types of improvement in the quality of one's writing can only be achieved at the expense of quantity—for instance, by spending more time on the editing of drafts. However, regardless of whether you use longhand or some form of dictation, there are several areas where quantity and quality go hand in hand, and where it can actually take less time to produce an effective piece of communication than an ineffective one. The six guidelines that follow take aim at that area.

Because writing problems vary in kind and degree from person to person, you may find that some of these suggestions either don't apply to you or appear self-evident. However, "fast, functional" writing skills are so important to most engineers and technical managers that even if you only find one or two of the suggestions really useful, this should more than repay you for your reading time.

1. Know what your audience expects

A vice-president of operations was making his annual tour of the company's outlying facilities. At one plant, he chatted with the assistant manager about the upcoming labor negotiations with the local union.

"I wish you'd send me a fairly detailed report of those negotiations," asked the vice-president. Flattered by this request, the assistant plant manager resolved to do an outstanding job of reporting. By the time negotiations were over, he had more than 50 pages of handwritten notes, which he decided to amplify with background information on some of the points the union or the management team had raised.

It took him two weeks after the close of negotiations to organize this material into a rough handwritten draft. Because of a secretarial bottleneck, it took him two more weeks to get this draft typed and edited, and then another two weeks to get the massive document typed up in final form. Nevertheless, as he signed the letter of transmittal in which he apologized for imperfections in the typing and commented on the shortage of good secretarial help, he felt rather proud of himself: The report read like a courtroom drama, and had a great degree of polish.

Unfortunately, before the report could even reach headquarters, the vice-president had made an acid phone call to the plant manager. Apparently, what the vice-president had really wanted was brief, day-by-day reports that he could review while negotiations were still in progress, and that he could pass on to another plant that started negotiations a week later. When he finally got the assistant plant manager's monumental opus, a lot of the information was old hat, the key points having been already passed along by the plant manager in a lengthy, long-distance phone call just before the close of negotiations.

Next year, the plant manager decided to handle the job of reporting himself; at the close of each day's session, he dictated his notes into a recorder—his secretary would type them up on the next day, so that he could ink in some comments and send them out as soon as he finished that day's negotiating. Since he didn't use a draft, his daily reports were not particularly polished; the inked-in comments and corrections looked rather informal—and yet these reports were acclaimed by the vice-president.

Moral of the story: Make sure you know what the audience wants. Don't be afraid to ask. There are times when speed, rather than quality or quantity, may be of the essence. In the example above, which is based on an actual situation, the assistant plant manager could have saved himself hours of toil by finding out what sort of details the vice-president did or did not want to know about, and whether quick dictation without a draft was acceptable.

Failure to find out or to really understand the audience's expectations is perhaps the single biggest time-waster in technical writing. An example of such inefficiency involves the engineer who uses exactly the same writing approach regardless of whether his report is primarily intended for the departmental archives, his immediate supervisor, or a financial executive on the appropriations committee.

If the report is just intended for a supervisor, for instance, there is nothing wrong with using technical jargon, and for keeping background information very brief. But if the report is to go to a financial executive—particularly one who lacks any sort of technical background—the engineer must avoid technical jargon, and explain some things that he would not need to explain to his boss, while leaving out technical details. In such a report, it is particularly important to make the first page or two tell the bulk of the story, in terms that are meaningful to the administrator.

If you are writing a dual-function report—say, one intended for both a financial executive and your boss—by all means get squared away with the latter on how the two approaches can be reconciled (e.g., by eventually giving your boss additional sections or informal notes that are omitted from the administrator's copy, by coming up with a preliminary version of the first two pages that you and your boss can review jointly before finalization, etc.).

When writing for publication, rather than for internal use, finding out the audience's expectations can be an equally great timesaver. Many magazines have booklets that discuss their "expectations" in regard to such things as writing style, quality of drawings, compatibility with readers' interests[1]—and are further prepared to conserve the author's time by reviewing outlines in some detail so as to minimize the need for revision of the final manuscript.

2. Mobilize your subconscious

Legend has it that on the day the opera "The Magic Flute" was scheduled to get its first performance, someone reminded Mozart that there still wasn't any overture, whereupon the composer calmly sat down and dashed off the magnificent overture just in time to give the score to the orchestra before the curtain went up. Actually, I think if this legend is true, the chances are that Mozart must have "precomposed" the overture in his mind days before he set it down on paper, so that leaving the actual writing until just before the deadline may not have been as risky as it seemed to others at the time.

Unfortunately, many of us tend to display a Mozart-like confidence in being able to meet a last-minute deadline regardless of whether we have done any precomposing or not.

If you have a writing project with a far-off deadline, and you don't feel like starting the actual writing right away, then don't. But I would still suggest that you draw up a rough outline as soon as you can, see how much information you already have available and how much more you will have to dig out, decide when you should start this digging out so that the information will be available at the proper time, and discuss the outline for the project with your boss or with a colleague. That way, you will not only get a feel for the emphasis needed (as discussed in the previous section), but you will be giving your subconscious mind a chance to come up with ideas while the project is incubating.

To come up with useful ideas, your subconscious should have some sort of framework—that way, perhaps some of the framework will be filled in for you while you shave in the morning. But if you wait till the last minute, even to prepare an outline or think about an approach, all your subconscious can do is to provide you with a feeling of anxiety.

Another constructive use for the prewriting period is to do some preselling. People hate shocks—which is what may happen when your recommendations based on a full year's work are coldly dropped on your department head's desk. It is fairer to him, and kinder to you and your writing time, to use the "let's let him in on part of the secret" technique. So, give your boss some inkling of what your report is going to recommend. Don't try to do a complete job of selling, but do give him a chance to get some exposure to your ideas. From his initial comments, you may also be able to obtain pointers on how to best link your ideas to his, and to the current objectives of the department. This can save a great deal of writing time in the long run, and will often let you get the constructive involvement of the reader much sooner than if you use the "This may come as a shock, but . . ." approach.

3. Build up and exploit your momentum

"I sit down with a fountain pen and paper and the story pours out. However lousy a section is, I let it go. I write on to the end. Then the subconscious mind has done what it can . . . The rest is simple effort . . . going over a chapter time and time again until, though you know it isn't right, it is the best you can do." That's how W. Somerset Maugham said he got his thoughts down on paper. If you have trouble getting started and building up momentum on a writing project, then the Maugham approach of putting down your thoughts in any old undisciplined way and eventually going back to polish them up, has a lot to recommend it. . . .

What do you do if you still don't succeed in unplugging your thoughts, regardless of whether you are using pen, pencil, stenographer or dictating machine? Here are a few additional pointers:

• Go over your outline . . . and pretend a good friend asked you to explain each point to him. If he were to ask questions such as "Why don't we just stick with the old process?" or "How does this gimmick work?," the chances are you wouldn't be at a loss for words—you would just say it like it is. So, pretend your friend is asking you questions that correspond to your outline, and write it like it is.

• Pretend that this is an examination, and that you have five minutes to get something down on paper for each thought on your outline—i.e., that if after five minutes you have nothing down on paper on the first point, you have nothing that you can be graded on, and you would thus get zero. (A couple of years ago, I took an aptitude test in which I was given a cast of two or three characters and had to shape them into a plot for a short story within ten minutes; then stop and shape another cast of characters into a new plot; then do this a few more times. Not considering myself a really fast writer, I was amazed at how much I could get down in ten minutes when the pressure was on, and when I knew I would be graded on content rather than style. It occurred to me that if I could pretend that I was in a similar situation when starting on a writing job, it would get me going—and it works!)

• Pick the best time to get started on a tough writing project. For most people, this may be the very beginning of the day, before things have come up to distract them. Others work best right after lunch. Still others find that the best time to unlock their thoughts is after dinner at home, and that once they get started that way, they can continue the project at work without much difficulty. If you haven't already done so, try to find a time pattern in the daily periods during which you feel sharpest and loosest; then exploit these periods to get started on tough projects.

Of course, getting started does not necessarily mean writing a brilliant beginning; it means getting started on the "meat" of the project. Very often a meaty fact-filled beginning is the most brilliant one anyway. But if you do feel the urge to think up something particularly striking or original, you may be better off doing this after most of the project is down on paper, rather than letting it be an initial stumbling block.

Once you have built up your momentum, how do you keep it?

First of all, try not to stop. Let someone take your phone messages. If you see a chance to finish the project by working late, do so. Resist the temptation to take a break until you start slowing down. (Personally, I find that sipping coffee or puffing on a cigar while doing difficult writing is preferable to running the risk of getting sidetracked by taking an unnecessary break.) Putting it into industrial engineering terminology, make sure you have a long enough production run to justify the setup time.

If you can't finish the project by working late, then it is usually better to leave it in the middle of a paragraph than to try and finish the particular section or thought. Finishing the thought the next day will be relatively easy, and this will serve to prime the pump for the next thought. I used to make the mistake of staying up until whatever hour was necessary to get a major section finished, only to find that it would take me all the next morning to stop resting on my laurels and to uncork my thoughts for the next section. Conversely, when I don't try so hard to come to a logical stopping place, I can usually continue the section without any difficulty the next morning and then launch right into the next section. (I can't take full credit for this approach; someone named Hemingway recommended it in a book called "A Movable Feast.")

4. Watch out for time-wasting verbosity

Not everybody has the uncoiling problems dealt with in the previous section. In fact, some people who can uncoil most of the time at the drop of a pencil encounter the opposite type of problem: How to avoid the verbosity that wastes their own time as well as that of the reader, so that they can make the most of their fluency as writers.

The first section in the 1966 "Efficient, Effective Writing" report[2] gave good advice on ways of recognizing and avoiding destructive verbosity. Although that section was aimed at making writing more informative, most of the suggestions can also lead to faster writing. For instance, the writer can usually save time by cutting down on passive verbs, abstract nouns, and prepositional phrases. (Obviously, it would have taken me longer to write the previous sentence had I said "Time can be saved by the writer if proper consideration is given to a reduction in the use of passive verbs, etc.")

Here are a few additional suggestions:

• Don't be legalistic in a nonlegal piece of writing. Lawyers have to be extremely careful to cover every possible contingency—hence they may feel justified in using strings of words that may differ only very slightly in meaning (e.g., null, void, of no legal force, etc.). But engineers are seldom justified in coming up with a phrase like "The *development, establishment,* and *implementation* of process-control *philosophies* and *policies* must take place at an *early* or *incipient* stage in a project." There just isn't enough difference between the italicized words to warrant using more than one of each; if you say "philosophies," 99 out of 100 readers will assume that this includes "policies."

• Don't waste time on excess hedging. For instance, when you say "Based on bench-scale experiments run at ten different temperatures, 300 F. produces the best yield," you have clearly indicated on what you base your conclusion, and there is no need to add hedge phrases such as ". . . 300 F. may tend to produce the best yields, assuming that these bench-scale results are freely applicable to commercial conditions . . ." Putting it another way, readers generally realize that almost any technical, business or philosophical statement that can be made is subject to limitations and qualifications; if you insist on pointing out limitations that are either obvious or unimportant, you are wasting time (and probably producing dull writing). Particularly wasteful and deadly is the "triple hedge"—e.g., ". . . may, under some circumstances, tend to . . ."

• Use more tables, illustrations and in-text listings to cut down on verbosity. . . . An item-by-item approach can result in tighter writing of all kinds, including the body of the report. (At the risk of appearing to disregard my own "hedging" caveat, I should point out that itemization can be carried to dull extremes, particularly once the triple sub-indentation stage is reached. But if you haven't been doing much itemizing, the technique can save you a lot of words in reducing generalities to specifics.)

• Consider leaving out some details altogether. Resist the temptation to tell the reader everything you know about the subject—just tell him what you think he needs to know. The latter is quite different from just telling him what you think he would like to hear—or from using a dual standard whereby you exclude all unfavorable details while including all those that are favorable. The memo or report that burdens the reader with every conceivable detail has not only taken the writer much more time than necessary, but suggests that the writer was unsure of himself because he left it to the reader to decide which of the details were significant. (If you are afraid that a short report won't adequately reflect the work that went into a particular study, you can always indicate the *type* of additional data that can be supplied on request.)

Of course, the classic example of detail elimination involves the man who got a letter from his landlord asking if he intended to vacate his apartment. The answer consisted of "Dear Sir: I remain, Yours Truly, Henry Smith."

Unfortunately, it's not always that easy to combine succinctness with politeness in handling one's correspondence, particularly when one is dictating it. Things that can waste the dictator's (and the reader's) time include:

• Marathon sentences. An occasional long sentence can supply useful variety, and may be needed in order to relate ideas to each other, but strings of runaway sentences impede efficient communication.

• Ditto for marathon paragraphs.

• Shotgun or overkill attacks on the topic, whereby the dictator, not being sure he has said what he really wanted to say, goes on to re-attack the topic in several other, equally roundabout, ways.

• Tendencies to throw in cliches and meaningless phrases just to keep the dictation process going. It is less wasteful to keep your secretary waiting for a minute while you find the right phrase than to dictate a phrase that adds nothing.

5. Save time via the "example" technique

Most of us find the going slow when we write on an abstract, philosophic level for any length of time. True, the amount of such writing can be minimized by translating abstract concepts into dollars and cents, but this cannot always be done. For instance, if you are trying to change the attitudes of foremen towards some community or labor-relations problem, neither payout time nor discounted cash flow is going to be of much help to you. But this does not mean that your communications with the foremen must be entirely on an abstract philosophic level.

Good philosophy is hard to write. There are all sorts of pitfalls: failure to define terms, or to relate them to the reader's frame of mind, failure to avoid either oversimplification or obtuseness, etc. But fortunately, you can write about an abstract topic—e.g., a desired change in attitude—without staying on the abstract level for very long. The idea here is to translate the abstract into the specific by using analogies, miniature case histories (either actual or hypothetical), projections of what might happen if the status quo were maintained, etc.

Of course, this "example" technique is not limited to communications that deal with corporate philosophy but should also be used when generalities and abstract concepts crop up in technical writing. Here, a one-sentence "for instance" can often take the place of a much longer explanation; a liberal sprinkling of such sentences can keep the discussion on the ground, and can result in an easier-to-write yet more-informative communication than one in which all abstract technical concepts are explained or promoted in only theoretical terms.

6. Know thyself, and thy writing

The difficulty in many articles dealing with improving your writing is their generality; they presuppose that all writers have the same problems. This report has tried to be more specific—for instance, by showing how *less* discipline can be a timesaver for the engineer who has trouble uncoiling, whereas *more* discipline can be an eventual timesaver for someone whose problem is verbosity or poor organization rather than lack of writing fluency.

Where do you fall within these two extremes? Should your prime emphasis be on boosting your writing output or on boosting its quality? If you have difficulty answering that question, you can get help quite readily.

For example, the 1966 "Efficient, Effective Writing" report[2] supplied some simple tests whereby you can evaluate the informative quality of your writing. If your score is low, perhaps your prime goal should be to make your writing more informative. At the beginning, this may reduce your page output because you will be spending more time on editing and rewriting. However, if you try to learn from your editing, you will find that output will eventually increase along with quality. This means spending a few minutes reviewing an edited draft to see what type of corrections are prevalent, and to think about ways of minimizing the need for these types of corrections in subsequent writing.

Discuss your writing with your boss. Some bosses are reluctant to initiate such a discussion because they have found that it takes almost as much tact to constructively criticize a subordinate's writing as it would to constructively criticize his wife or family. But if you initiate such a discussion (I mean about your writing, not your wife), the chances are that overly defensive postures can be avoided, and that you and your boss will both gain a better insight into problem areas.

Ask your boss whether he thinks you are spending too much or too little time on writing. For instance, based on samples, would he settle for draftless dictation on some reports in order to give you more time for engineering?

If one of your writing problems is that interruptions are always breaking your train of thought, ask your boss about ways of getting more privacy, or about occasionally working at home to make headway on difficult writing chores.

In order to discuss other aspects of your writing, you may want to go over a sort of writing-inventory checklist, such as the one published in the *Harvard Business Review* article "What Do You Mean I Can't Write?"[3] This can be a good starting point in getting your supervisor's views on such matters as whether you sometimes present too many opinions and not enough supporting data (or vice versa), whether you tend to under- or overestimate his familiarity with your work, etc.

A concluding thought: The engineer who progresses in his company tends to do more writing every year. Eventually, he may also have to supervise the writing of subordinates; many of their reports and memos will go out over his signature and will reflect on the caliber of the work done under his supervision. Thus the engineer who becomes adept at fast, functional writing is in an enviable position—he can save a significant and increasing amount of time, establish his credentials as an efficient communicator, and eventually help his subordinates solve their own communication problems. We hope that the suggestions in various sections of this report will help point the way, and will let you apply the "work smarter, not harder" principle to your communicating.

Acknowledgement
Some of the ideas in this final section stem from various members of *Chemical Engineering*'s editorial staff, and also from Manny Meyers of Picatinny Arsenal, and Peter J. Rankin of Basford Inc.

References
1. See, for instance, "How to Write for *Chemical Engineering*," a booklet available from *Chemical Engineering*, 330 W. 42 St., New York City.
2. Johnson, Thomas P., How Well Do You Inform?, *Chem. Eng.*, Mar. 14, 1966 (Reprint No. 295).
3. Fielden, J., *Harvard Bus. Rev.*, May–June 1964, p. 147.

Using PAFEO planning

JOHN KEENAN

A former technical writer for SmithKline, John Keenan is now Chairman of the Department of English and Communication Arts at La Salle University in Philadelphia.

Surveys tell us that "lack of clarity" is the most frequently cited weakness when executives evaluate their employees' writing. But since lack of clarity results from various bad habits, it is often used as a convenient catchall criticism. Pressed to explain what they meant by lack of clarity, readers would probably come up with comments such as these:

"I had to read the damn thing three times before I got what he was driving at. At least I *think* I got his point; I'm not sure."

"I couldn't follow her line of thought."

"I'm just too busy to wade through a lot of self-serving explanation and justification. If he has something to say, why didn't he just say it without all that buildup?"

Most unclear writing results from unclear thinking. But it is also true that habits of clear writing help one to think clearly. When a writer gets thoughts in order and defines the purpose and goal, what he wants to say may no longer be exactly the same as it was originally. Sometimes the writer may even decide it is better *not* to write. Whatever the case, the work done *before* the first draft in the prewriting process and *after* the draft in the revising process is often the difference between successful communication and time-wasting confusion. . . .

Think PAFEO

Some years ago I made up a nonsense word to help my students remember the important steps in constructing a piece of writing. It seemed to stick in their minds. In fact, I keep running into former students who can't remember my name, but remember PAFEO because it proved to be so useful. Here are the magic ingredients.

P stands for purpose
A stands for audience
F stands for format
E stands for evidence
O stands for organization

Put together, they spell PAFEO, and it means the world to me!

Purpose

Put the question to yourself, "Why am I writing this?" ("Because I have to," is not a sufficient answer.) This writing you're going to do must aim at accomplishing something; you must be seeking a particular response from your reader. Suppose, for example, you're being a good citizen and writing to your congressman on a matter of interest to you. Your letter may encompass several purposes. You are writing to inform him of your thinking and perhaps to persuade him to vote for or against a certain bill. In explaining your position, you may find it useful to narrate a personal experience or to describe a situation vividly enough to engage his emotions. Your letter may therefore include the four main kinds of writing—exposition, argument, narration, and description—but one of them is likely to be primary. Your other purposes would be subordinated to your effort to persuade the congressman to vote your way. So your precise purpose might be stated: "I am writing to Congressman Green to persuade him to vote for increased social security benefits provided by House Bill 5883."

Write it out in a sentence just that way. Pin it down. And make it as precise as you can. You will save yourself a good deal of grief in the writing process by getting as clear a focus as you can on your purpose. You'll know better what to include and what to omit. The many choices that combine to form the writing process will be made easier because you have taken the time and the thought to determine the exact reason you have for writing this communication.

When you're writing something longer than a letter, like a report, you can make things easier for yourself and your reader by making a clear statement of the central idea you're trying to develop. This is your thesis. To be most helpful, it must be a complete sentence, not just the subject of a sentence. "Disappointing sales" may be the subject you are writing about, but it is not a thesis. Stating it that way won't help you organize the information in your report. But if you really say something specific about "disappointing sales," you will have a thesis that will provide a framework for development. For example:

Disappointing sales may be attributed to insufficient advertising, poor selection of merchandise, and inadequate staffing during peak shopping hours.

The frame for the rest of the report now exists. By the time you have finished explaining how each cause contributes to the result (disappointing sales), the report is finished. Not only finished, but clear; it follows an orderly pattern. The arguments, details, illustrations, statistics, or quotations are relevant because the thesis has given you a way of keeping your purpose clearly in focus.

Of course the ideal method is to get your purpose and your thesis down before you write a first draft, but I can tell you from experience that it doesn't always work that way. Sometimes the only way to capture a thesis statement is to sneak up on it by means of a meandering first draft. When you've finished that chore, read the whole thing over and try to ambush the thesis by saying, "OK, now just what am I trying to say?" Then *say it* as quickly and concisely as you can—out loud, so you can hear how it sounds. Then try to write it down in the most precise and concise way you can. Even if you don't capture the exact thesis statement you're looking for, there's a good chance that you will have clarified your purpose sufficiently to make your next revision better.

Audience

Most of the writing the average person does is aimed at a specific reader: the congressman, the boss, a business subordinate, a customer, a supplier. Having a limited audience can be an advantage. It enables the writer to analyze the reader and shape the writing so that it effectively achieves the purpose with that particular reader or group of readers.

Here are some questions that will help you communicate with the reader:

1. How much background do I need to give this reader, considering his or her position, attitude toward the subject, and experience with this subject?

2. What does the reader need to know, and how can I best give him or her this information?

3. How is my credibility with this reader? Must I build it gradually as I proceed, or can I assume that he or she will accept certain judgments based on my interpretations?

4. Is the reader likely to agree or disagree with my position? What tone would be most appropriate in view of this agreement or disagreement?

The last question, I think, is very important. It implies that the writer will try to see the reader's point of view, will bend every effort to look at the subject the way the reader will probably look at it. That isn't easy to do. Doing it takes both imagination and some understanding of psychology. But it is worth the considerable effort it involves; it is a gateway to true communication.

Let me briefly suggest a point here that is easily overlooked in the search for better techniques of communication. It is simply this: *better writing depends on better reading.* Technique isn't everything. One cannot learn to be imaginative and understanding in ten easy lessons, but reading imaginative literature—fiction, poetry, drama—is one way of developing the ability to put yourself in someone else's place so you can see how the problem looks from that person's angle. The writer who limits reading to his or her own technical field is building walls around the imagination. The unique value of imaginative literature is the escape that it allows from the prison of the self into the experiences and emotions of persons of different backgrounds. That ability to understand why a reader is likely to think and act in a certain way is constantly useful to the writer in the struggle to communicate.

Format

Having thought about your purpose and your audience, consider carefully the appropriate format for this particular communication you are writing. Of course the range of choice may be limited by company procedures, but even the usual business formats allow some room for the writer to use ingenuity and intelligence.

The key that unlocks this ingenuity is making the format as well as the words work toward achieving your purpose. You can call attention to important points by the way you arrange your material on the page.

Writers sometimes neglect this opportunity, turning the draft over to the typist without taking the trouble to plan the finished page. Remember the reader, his or her desk piled high with things to be read, and try to make your ideas as clear as possible on the page as the reader glances at it.

External signals such as headings, underlining, and numerals are ways of giving the reader a quick preview. Use them when they are appropriate to your material.

Don't be afraid of using white space as a way of drawing attention to key ideas.

To see how experts use format for clarity and emphasis, take a look at some of the sales letters all of us receive daily. Note how these persuasive messages draw the reader's attention to each advantage of the product. You may not want to go that far, but the principle of using the format to save the reader's time is worth your careful consideration.

Overburdened executives are always looking for shortcuts through the sea of paper. Many admit that they cannot possibly read every report or memo. They scan, they skip, they look for summaries. A good format identifies the main points quickly and gives an idea of the organizational structure and content. If the format looks logical and interesting, readers may be lured into spending more time on your ideas. You've taken the trouble to separate the important from the unimportant, thus saving them time. The clear format holds promise of a clear analysis, and what boss can be too busy for that?

Evidence

Unless you're an authority on your subject, your opinions carry only as much weight as the evidence you can marshal to support them. The more evidence you can collect before writing, the easier the writing task will be. Evidence consists of the facts and information you gather in three ways: (1) through careful observation, (2) through intelligent fieldwork (talking to the appropriate persons), and (3) through library research.

When you follow the evidence where it leads and form a hypothesis, you are using inductive reasoning, the scientific method. If your evidence is adequate, representative, and related to the issue, your conclusion must still be considered a probability, not a certainty, since you can never possess or weigh *all* the evidence. At some point you make the "inductive leap" and conclude that the weight of evidence points to a theory as a probability. The probability is likely to be strong if you are aware of the rules of evidence.

The Rules of Evidence

Rule 1. Look at the evidence and follow where it leads.

The trick here is not to let your own bias seduce you into selecting only the evidence you agree with. If you aren't careful, you unconsciously start forcing the evidence to fit the design that seems to be emerging. When fact A and fact B both point toward the same conclusion, there is always the temptation to *make* fact C fit. Biographer Marchette Chute's warning is worth heeding: " . . . you will never succeed in getting at the truth if you think you know, ahead of time, what the truth ought to be."

A reliable generalization ought to be based on a number of verifiable, relevant facts—the more the better. Logicians tell us that evidence supporting a generalization must be (1) known or available, (2) sufficient, (3) relevant, and (4) representative. Let's see how these apply to the generalization, "Cigarette smoking is hazardous to your health."

The evidence accumulated in animal studies and in comparative studies of smokers and nonsmokers has been published in medical journals: it is therefore known and available. Is it sufficient? The government thinks so; the tobacco industry does not. Which is more likely to have a bias that might make it difficult to follow where the evidence leads?

Tobacco industry spokesmen argue that animal studies are not necessarily relevant to human beings and point to individuals of advanced age who have been smoking since childhood as living refutations of the supposed evidence. But the number of cases studied is now in the thousands, and studies have been done with representative samplings. The individuals who have smoked without apparent damage to their health are real enough, but can they be said to be numerous enough to be representative of the usual effects of tobacco on humans? "Proof" beyond doubt lies only in mathematics, but the evidence in this example is the kind on which a sound generalization can be based.

Rule 2. Look for the simplest explanation that accounts for all the evidence.

When the lights go out, the sudden darkness might be taken as evidence of a power failure. But a quick investigation turns up other evidence that must be accounted for: the streetlights are still on; the refrigerator is still functioning. So a simpler explanation may exist, and a check of the circuit breakers or fuse box would be appropriate.

Rule 3. Look at all likely alternatives.

Likely alternatives in the example just discussed would include such things as burned-out bulbs, loose plugs, and defective outlets.

Rule 4. Beware of absolute statements.

In the complexity of the real world, it is seldom possible to marshal sufficient evidence to permit an absolute generalization. Be wary of writing general statements using words like *all*, *never* or *always*. Sometimes these words are implied rather than stated, as in this example:

> Jogging is good exercise for both men and women.
> [Because it is unqualified, the statement means *all* men and women. Since jogging is not good for people with certain health problems, this statement would be better if it said *most* or *many* men and women.]

Still, caution is always necessary. Induction has its limitations, and a hypothesis is best considered a probability subject to change on the basis of new evidence.

The other kind of reasoning we do is called deduction. Instead of starting with particulars and arriving at a generalization, deduction starts with a general premise or set of premises and works toward the conclusion necessarily implied by them. If the premise is true, it follows that the conclusion must be true. This logical relationship is called an inference. A fallacy is an erroneous or unjustified inference. When you are reasoning deductively, you can get into trouble if your premise is faulty or if the route from premise to conclusion contains a fallacy. Keep the following two basic principles in mind, and watch out for the common fallacies that can act as booby traps along the path from premise to conclusion.

Two Principles for Sound Deduction

1. The ideas must be true; that is, they must be based on facts that are known, sufficient, relevant, and representative.

2. The two ideas must have a strong logical connection.

If your inductive reasoning has been sound enough to take care of the first principle, let me warn you that the second is not so simple as it sounds. We have all been exposed to so much propaganda and advertising that distort this principle that we may have trouble recognizing the weak or illogical connection. Nothing is easier than to tumble into one of the commoner logic fallacies. As a matter of fact, no barroom argument would be complete without one of these bits of twisted logic.

Twisted Logic

Begging the question. Trying to prove a point by repeating it in different words.

Women are the weaker sex because they are not as strong as men.

Our company is more successful because we outsell the competition.

Non sequitur. The conclusion does not follow logically.

I had my best sales year; the company's stock should be a big gainer this year.

I was shortchanged at the supermarket yesterday. You can't trust these young cashiers at all.

Post hoc. Because an event happened first, it is presumed to be the cause of the second event.

Elect the Democrats and you get war; elect the Republicans and you get a depression.

Oversimplification. Treating truth as an either-or proposition, without any degrees in between.

> Better dead than Red.
>
> America—love it or leave it.
>
> If we do not establish our sales leadership in New York, we might as well close up our East Coast outlets.

False analogy. Because of one or two similarities, two different things are assumed to be *entirely* similar.

> You can lead a horse to water but you can't make him drink; there is no sense in forcing kids to go to school.
>
> Giving the Canal back to Panama would be like giving the Great Lakes back to the Indians.

Seen in raw form, as in the examples, these fallacies seem easy to avoid. Yet all of us fall into them with dismaying regularity. A writer must be on guard against them whenever he or she attempts to draw conclusions from data.

Organization

Experienced writers often use index cards when they collect and organize information. Cards can be easily arranged and rearranged. By arranging them in piles, you can create an organizational plan. Here's how you play the game.

    1. You can write only one point on each card—one fact, one observation, one opinion, one statistic, one whatever.

    2. Arrange the cards into piles, putting all closely related points together. All evidence related to marketing goes in one pile, all evidence related to product development goes in another pile, and so on.

    3. Now you can move the piles around, putting them in sequence. What kind of sequence? Consider one of the following commonly used principles:

> **Chronological.** From past to present to future.
> Background, present status, prospects
>
> **Spatial.** By location.
> New England territory, Middle Atlantic, Southern
>
> **Logical.** Depends on the topic.
> Classification and division according to a consistent principle (divisions of a company classified according to function)
> Cause and effect (useful in troubleshooting manuals)

> **Problem-analysis-solution.** Description of problem, why it exists, what to do about it.
>
> **Order of importance.** From least important to most important or from most important to least important.

The choice of sequence will depend largely on the logic of the subject matter and the needs of your audience.

   4. Go through each pile and arrange the cards in an understandable sequence. Which points need to precede others in order to present a clear picture?

   5. That's it! You now have an outline that is both orderly and flexible. You can add, subtract, or rearrange whenever necessary.

Summary

Think PAFEO. Use this word to remind you to clarify your purpose and analyze your audience before you write. When you have a clear sense of purpose, create a thesis that will act as a frame for your ideas.

   Choose a format for your communication that will help the reader identify the main points quickly.

   Collect your evidence before you write by observing, interviewing, and doing library research. Use the index card system to help you organize your evidence. Keep in mind the rules of evidence before you draw conclusions. Be alert to the common logic fallacies that can easily undermine the deductive process.

   Try the file card system as a way of organizing your material. Write one point on each card and then place closely related points together in a pile. Place the piles in sequence according to a principle of organization you select. Some useful organizing principles include: chronological, spatial, logical, problem-analysis-solution, and order of importance. The sequence you choose will depend on the logic of the subject matter and the needs of your audience.

The writing process

MICHAEL E. ADELSTEIN

Michael E. Adelstein is Professor of English at the University of Kentucky and coauthor with W. Keats Sparrow of *Business Communications* (Harcourt Brace Jovanovich, 1983).

The writing process

Their prevalent belief in the myths of genius, inspiration, and correctness prevents many people from writing effectively. If these individuals could realize that they can learn to write, that they cannot wait to be inspired, and that they must focus on other aspects besides correctness, then they can overcome many obstacles confronting them. But they should be aware that writing, like many other forms of work, is a process. To accomplish it well, people should plan on completing each one of its five stages. The time allocations may vary with the deadline, subject, and purpose, but the work schedule should generally follow this pattern:

1. Worrying—15%
2. Planning—10%
3. Writing—25%
4. Revising—45%
5. Proofreading—5%

Note that only 25% of the time should be spent in writing; the rest—75%—should be spent in getting ready for the task and perfecting the initial effort. Observe also that more time is spent in revising than in any other stage, including writing.

Stage One—Worrying (15%)

Worrying is a more appropriate term for the first operation than *thinking* because it suggests more precisely what you must do. When you receive a writing assignment, you must avoid blocking it out of your mind until you're ready to write; instead, allow it to simmer while you try to cook up some ideas. As you stew about the subject while brushing your teeth, taking a shower, getting dressed, walking to class, or preparing for bed, jot down any pertinent thoughts. You need only note them on scrap paper or the back of an envelope. But if you fail to write them down, you'll forget them.

Another way to relieve your worrying is to read about the subject or discuss it with friends. The chances are that others have wrestled with similar problems and written their ideas. Why not benefit from their experience? But if you think that there is nothing in print about the subject or a related one— either because your assignment is restricted, localized, or topical—then you should at least discuss it with friends.

Let's say, for example, that as a member of the student government on your campus, you've been asked to investigate the possibility of opening a student-operated bookstore to combat high textbook prices. During free hours, you might drop into the library to read about the retail book business and, if possible, about college bookstores. If your student government is affiliated with the National Student Association, you might write to it for information about policies and practices on other campuses. In addition, you might start talking about the problem to the manager of your college bookstore, owners of other bookstores in your community, student transfers from other schools, managers of college bookstores in nearby communities, and some of your professors. Because you might be concerned about the public-relations reper-cussions of a student bookstore, you should propose the venture to college officials. At this point, you will have heard many different ideas. To discuss some of them, bring up the subject of a student bookstore with friends. As you talk, you will be clarifying and organizing your thoughts.

The same technique can be applied to less involved subjects, such as a description of an ideal summer job, or an explanation of a technical matter like input-output analysis. By worrying about the subject as you proceed through your daily routine, by striving to get ideas from books and people or both, and by clarifying your thoughts in conversation with friends, you will be ready for the next step. Remember: you can proceed only if you have some ideas. You must have something to write about.

Stage Two—Planning (10%)

Planning is another term for organizing or for the task that students dread so— *outlining.* You should already have had some experience with this process. Like many students, unfortunately, you may have learned only how to outline a paper after writing it. This practice is a waste of time. If you want to operate efficiently, then it follows that you will have to specialize in each of the writing stages in turn. When you force yourself simultaneously to conceive of ideas, to arrange them in logical order, and to express them in words, you cannot per-form all of these complex activities as effectively as if you had concentrated on each one separately. By dividing the writing process into five stages, and by focusing on each one individually, the odds are that the result will be much better than you had thought possible. Of course you will hear about people who never plan their writing but who do well anyway, just as you may know someone who skis or plays golf well without having had a lesson. There are always some naturals who can flaunt the rules. But perhaps these people could have significantly improved their writing, golf, or skiing if they had received some formal instruction and proceeded in the prescribed manner.

An efficient person plans his work. Many executives take a few minutes before leaving the office or retiring at night to jot down problems to attend to the following day. Many vacationers list tasks to do before leaving home: stop the newspaper, turn down the refrigerator, inform the mailman, cut off the hot water, check the car, mail the mortgage payment, notify the police, and the like. Of course, you can go on vacation without planning—just as you can write without planning—but the chances are that the more carefully you organize, the better the results will be.

In writing, planning consists mainly in examining all your ideas, eliminating the irrelevant ones, and arranging the others in a clear, logical order. Whether you write a formal outline or merely jot down your ideas on scrap paper is up to you. The outline is for your benefit; follow the procedure that helps you the most. But realize that the harder you work on perfecting your outline, the easier the writing will be. As you write, therefore, you can concentrate on formulating sentences instead of also being concerned with thinking of ideas and trying to organize them. For the present, realize the importance of planning. . . .

Stage Three—Writing (25%)

When you know what you want to say and have planned how to say it, then you are ready to write. This third step in the writing process is self-explanatory. You need only dash away your thoughts, using pen, pencil, or typewriter, whichever you prefer. Let your mind flow along the outline. Don't pause to check spelling, worry about punctuation, or search for exact words. If new thoughts pop into your mind, as they often do, check them with your outline, and work them into your paper if they are relevant. Otherwise discard them. Keep going until either the end or fatigue halts you. And don't worry if you write slowly: many people do.

Stage Four—Revising (45%)

Few authors are so talented that they can express themselves clearly and effectively in a first draft. Most know that they must revise their papers extensively. So they roll up their sleeves and begin slashing away, cutting out excess verbiage, tearing into fuzzy sentences, stabbing at structural weaknesses, and knifing into obfuscation. Revision is painful: removing pet phrases and savory sentences is like getting rid of cherished possessions. Just as we dislike to discard old magazines, books, shoes, and clothes, so we hate to get rid of phrases and sentences. But no matter how distasteful the process may be, professional writers know that revising is crucial; it is usually the difference between a mediocre paper and an excellent one.

Few students revise their work well; they fail to realize the importance of this process, lack the necessary zeal, and are unaware of how to proceed. Once you understand why you seem to be naturally adverse to revision, you may overcome your resistance.

Because writing requires concentrated thought, it is about the most enervating work that humans perform. Consequently, upon completing a first draft, we are so relieved at having produced something that we tidy it up slightly, copy it quickly, and get rid of it immediately. After all, why bother with it anymore? But there's the rub. The experienced writer knows that he has to sweat through it again and again, looking for trouble, willing to rework cumbersome passages, and striving always to find a better word, a more felicitous phrase, and a smoother sentence. John Galbraith, for example, regularly writes five drafts, the last one reserved for inserting "that note of spontaneity" that ironically makes his work appear natural and effortless. The record for revision is probably held by Ernest Hemingway, who wrote the last page of *A Farewell to Arms* thirty-nine times before he was satisfied!

To revise effectively, you must not only change your attitude, but also your perspective. Usually we reread a paper from our own viewpoint, feeling proud of our accomplishment. We admire our cleverness, enjoy our graceful flights, and glow at our fanciful turns. How easy it is to delude ourselves! The experienced writer reads his draft objectively, looking at it from the standpoint of his reader. To accomplish this, he gets away from the paper for a while, usually leaving it until the following morning. You may not be able to budget your time this ideally, but you can put the paper aside while you visit a friend, grab a bite to eat, or phone someone. Unless you divorce yourself from the paper, you will probably be under its spell; that is, you will reread it with your mind rather than with your eyes. You will see only what you think is on the page instead of what is actually there. And you will not be able to transport yourself from your role of writer to that of critic.

Only by attacking your paper from the viewpoint of another person can you revise it effectively. You must be anxious to find fault and you must be honest with yourself. If you cannot or will not realize the weaknesses in your writing, then you cannot correct them. This textbook should open your eyes to many things that can go wrong, but above all you must *want* to find them. If you blind yourself to the bad, then your work will never be good. You must convince yourself that good writing depends on good rewriting. This point cannot be repeated too frequently or emphasized too much. Tolstoy revised

War and Peace—one of the world's longest and greatest novels—five times. The brilliant French stylist Flaubert struggled for hours, even days, trying to perfect a single sentence. But question your professors about their own writing. Many—like me—revise papers three, four, or five times before submitting them for publication. You may not have this opportunity, your deadlines in school and business may not permit this luxury; but you can develop a respect for revision and you can devote more time and effort to this crucial stage in the writing process. . . .

Stage Five—Proofreading (5%)

Worrying, planning, writing, and revising are not sufficient. The final task, although not as taxing as the others, is just as vital. Proofreading is like a quick check in the mirror before leaving for a date. A sloppy appearance can spoil a favorable acceptance of you or your paper. If, through your fault or your typist's, words are missing or repeated, letters are transposed, or sentences juxtaposed, a reader may be perturbed enough to ignore or resist your ideas or information. Poor proofreading of an application letter can cost you a job. Such penalties may seem unfair, but we all react in this fashion. What would you think of a doctor with dirty hands? His carelessness would not only disgust you but would also raise questions about his professional competence. Similarly, carelessness in your writing antagonizes readers and raises questions in their minds about your competence. The merit of a person or paper should not depend on appearances, but frequently it does. Being aware of this possibility, you should keep yourself and your writing well groomed.

Poor proofreading results from failure to realize its importance, from inadequate time, and from improper effort. Unless you are convinced that scrutinizing your final copy is important, you cannot proofread effectively. If you realize the significance of this fifth step in the writing process, then you will not only have the incentive to work hard at proofreading but also will allow time for it in planning.

Like most things, proofreading takes time. We usually run out of time for things that we don't care about. But there's always time for what we enjoy or what is important to us. So with proofreading; lack of time is seldom a legitimate excuse for doing it badly.

But lack of technique is. To proofread well, you must forget the ideas in a paper and focus on words. Slow down your reading speed, stare hard at the black print, and search for trouble rather than trying to finish quickly. . . . Some professional proofreaders start with the last sentence and read backward to the first. Whatever technique you adopt, work painstakingly, so that a few careless errors will not spoil your efforts. If you realize the importance of proofreading, view it as one of the stages in the writing process, and labor at it conscientiously, your paper will reflect your care and concern. . . .

Each stage is important

These then are the five stages in the writing process: worrying, planning, writing, revising, and proofreading. Each is vital. You need to become proficient at each of them, although, at some later point in your life, an editor or efficient secretary may relieve you of some proofreading chores. But in your career as a student and in your early business years, you will be on your own to guide your paper through the five stages. Since it is your paper, no one else will be as interested in it as you. It's your offspring—to nourish, to cherish, to coddle, to bring up, and finally, to turn over to someone else. You will be proud of it only if you have done your best throughout its growth and development.

The direct writing process for getting words on paper

PETER ELBOW

Peter Elbow teaches at the State University of New York at Stony Brook. He is also the author of *Writing without Teachers* (Oxford, 1973), an innovative introduction to the writing process.

*T*he direct writing process is most useful if you don't have much time or if you have plenty to say about your topic. It's a kind of let's-get-this-thing-over-with writing process. I think of it for tasks like memos, reports, somewhat difficult letters, or essays where I don't want to engage in much new thinking. It's also a good approach if you are inexperienced or nervous about writing because it is simple. . . .

Unfortunately, its most common use will be for those situations that aren't supposed to happen but do: when you have to write something you *don't* yet understand, but you also don't have much time. The direct writing process may not always lead to a satisfactory piece of writing when you are in this fix, but it's the best approach I know.

The process is very simple. Just divide your available time in half. The first half is for fast writing without worrying about organization, language, correctness, or precision. The second half is for revising.

Start off by thinking carefully about the audience (if there is one) and the purpose for this piece of writing. Doing so may help you figure out exactly what you need to say. But if it doesn't, then let yourself put them out of mind. You may find that you get the most benefit from ignoring your audience and purpose at this early stage of the writing process. . . .

In any event spend the first half of your time making yourself write down everything you can think of that might belong or pertain to your writing task: incidents that come to mind for your story, images for your poem, ideas and facts for your essay or report. Write fast. Don't waste any time or energy on how to organize it, what to start with, paragraphing, wording, spelling, grammar, or any other matters of presentation. Just get things down helter-skelter.

If you can't find the right word just leave a blank. If you can't say it the way you want to say it, say it the wrong way. (If it makes you feel better, put a wavy line under those wrong bits to remind you to fix them.)*

I'm not saying you must never pause in this writing. No need to make this a frantic process. Sometimes it is very fruitful to pause and return in your mind to some productive feeling or idea that you've lost. But don't stop to worry or criticize or correct what you've already written.

While doing this helter-skelter writing, don't allow too much digression. Follow your pencil where it leads, but when you suddenly realize, "Hey, this has nothing to do with what I want to write about," just stop, drop the whole thing, skip a line or two, and get yourself back onto some aspect of the topic or theme.

Similarly, don't allow too much repetition. As you write quickly, you may sometimes find yourself coming back to something you've already treated. Perhaps you are saying it better or in a better context the second or third time. But once you realize you've done it before, stop and go on to something else.

When you are trying to put down everything quickly, it often happens that a new or tangentially related thought comes to mind while you are just in the middle of some train of thought. Sometimes two or three new thoughts crowd in on you. This can be confusing: you don't want to interrupt what you are on, but you fear you'll forget the intruding thoughts if you don't write them down. I've found it helpful to note them without spending much time on them. I stop right at the moment they arrive—wherever I am in my writing—and jot down a couple of words or phrases to remind me of them, and then I continue on with what I am writing. Sometimes I jot the reminder on a separate piece of paper. When I write at the typewriter I often just put the reminder in caps inside double parentheses ((LIKE THIS)) in the middle of my sentence. Or I simply start a new line
LIKE THIS
and then start another new line to continue my old train of thought. But sometimes the intruding idea seems so important or fragile that I really want to go to work on it right away so I don't lose it. If so, I drop what I'm engaged in and start working on the new item. I know I can later recapture the original thought because I've already written part of it. The important point here is that what you produce during this first half of the writing cycle can be very fragmented and incoherent without any damage at all.

There is a small detail about the physical process of writing down words that I have found important. Gradually I have learned not to stop and cross out something I've just written when I change my mind. I just leave it there and

---

* An excerpt from a letter giving me feedback on an earlier draft: "I tried the direct writing process. Though it sounds simple enough, I . . . see now that in the past I've often interrupted the flow of writing by spending disproportionate time on spelling, punctuation, etc. I can spend hours on an opening paragraph stroking the words to death; then, if there's a deadline, have to rush through the remainder." (Joanne Turpin, 7/24/78.)

write my new word or phrase on a new line. So my page is likely to have lots of passages that look like
Many of my pages
Still I don't mean that you should stop and rewrite every passage till you are happy with it.
This kind of appearance.

What is involved here is developing an increased tolerance for letting mistakes show. If you find yourself crumpling up your sheet of paper and throwing it away and starting with a new one every time you change your mind, you are really saying, "I must destroy all evidence of mistakes." Not quite so extreme is the person who scribbles over every mistake so avidly that not even the tail of the "y" is visible. Stopping to cross out mistakes doesn't just waste psychic energy, it distracts you from full concentration on what you are trying to say.

What's more, I've found that leaving mistakes uncrossed out somehow makes it easier for me to revise. When I cross out all my mistakes I end up with *a draft*. And a draft is hard to revise because it is a complete whole. But when I leave my first choices there littering my page along with some second and third choices, I don't have a draft, I just have a succession of ingredients. Often it is easier to whip that succession of ingredients into something usuable than, as it were, to *undo* that completed draft and turn it into a better draft. It turns out I can just trundle through that pile of ingredients, slash out some words and sections, rearrange some bits, and end up with something quite usable. And quite often I discover in retrospect that my original "mistaken phrase" is really better than what I replaced it with: more lively or closer to what I end up saying.

● ● ●

If you only have half an hour to write a memo, you have now forced yourself in fifteen minutes to cram down every hunch, insight, and train of thought that you think might belong in it. If you have only this evening to write a substantial report or paper, it is now 10:30 P.M., you have used up two or two and a half hours putting down as much as you can, and you only have two more hours to give to this thing. You must stop your raw writing now, even if you feel frustrated at not having written enough or figured out yet exactly what you mean to say. If you started out with no real understanding of your topic, you certainly won't feel satisfied with what is probably a complete mess at this point. You'll just have to accept the fact that of course you will do a poor job compared to what you could have done if you'd started yesterday. But what's more to the point now is to recognize that you'll do an even crummier job if you steal any of your revising time for more raw writing. Besides, you will have an opportunity during the revising process to figure out what you want to say—what all these ingredients add up to—and to add a few missing pieces. It's important to note that when I talk about revising in this book I mean something much more substantial than just tidying up your sentences.

So if your total time is half gone, stop now no matter how frustrated you are and change to the revising process. That means changing gears into an entirely different consciousness. You must transform yourself from a fast-and-loose-thinking person who is open to every whim and feeling into a ruthless, toughminded, rigorously logical editor. Since you are working under time pressure, you will probably use quick revising or cut-and-paste revising. . . .

• • •

Direct writing and quick revising are probably good processes to start with if you have an especially hard time writing. They help you prove to yourself that you *can* get things written quickly and acceptably. The results may not be the very best you can do, but they work, they get you by. Once you've proved you can get the job done you will be more willing to use other processes for getting words down on paper and for revising—processes that make greater demands on your time and energy and emotions. And if writing is usually a great struggle, you have probably been thrown off balance many times by getting into too much chaos. The direct writing process is a way to allow a limited amount of chaos to occur in a very controlled fashion.

It's easiest to explain the direct writing process in terms of pragmatic writing: you are in a hurry, you know most of what you want to say, you aren't trying for much creativity or brilliance. But I also want to stress that the direct writing process can work well for very important pieces of writing and ones where you haven't yet worked out your thinking at all. But one condition is crucial: you must be confident that you'll have no trouble finding lots to say once you start writing. . . .

As I wrote many parts of this book, for example, I didn't have my thinking clear or worked out by any means, I couldn't have made an outline at gunpoint, and I cared deeply about the results. But I knew that there was lots of *stuff* there swirling around in my head ready to go down on paper. I used the direct process. I just wrote down everything that came to mind and went on to revise.

But if you want to use the direct writing process for important pieces of writing, you need plenty of time. You probably won't be able to get them the way you want them with just quick revising. You'll need thorough revising or revising with feedback. . . . For important writing I invariably spend more time revising than I do getting my thoughts down on paper the first time.

Main steps in the direct writing process

• If you have a deadline, divide your total available time: half for raw writing, half for revising.

• Bring to mind your audience and purpose in writing but then go on to ignore them if that helps your raw writing.

- Write down as quickly as you can everything you can think of that pertains to your topic or theme.

- Don't let yourself repeat or digress or get lost, but don't worry about the order of what you write, the wording, or about crossing out what you decide is wrong.

- Make sure you stop when your time is half gone and change to revising, even if you are not done.

- The direct writing process is most helpful when you don't have difficulty coming up with material or when you are working under a tight deadline.

Transform writer-based prose into reader-based prose

LINDA FLOWER

Linda Flower is Associate Professor of English at Carnegie-Mellon University and a recognized authority on the teaching of writing.

Good writers know how to transform writer-based prose (which works well for them) into reader-based prose (which works for their readers as well). Writing is inevitably a somewhat egocentric enterprise. We naturally tend to talk to ourselves when composing. As a result, we often need self-conscious strategies for trying to talk to our reader.

Strategy 1: Set up a shared goal

The first strategy for adapting a paper to a reader is to create a shared goal. Try to find a reason for writing your paper and a reason for reading it that both you and your reader share. (Remember that your desire to convey information will not necessarily be met by your reader's desire to receive it.) Then organize your ideas and your arguments around this common goal. You will need to consider the knowledge, attitudes, and needs of your reader, as discussed previously.

A shared goal can be a powerful tool for persuasion. To illustrate, try this exercise:

You have just been commissioned to write a short booklet on how to preserve older homes and buildings, which the City Historical Society wants to distribute throughout a historical section of the city in an effort to encourage preservation. Most of your readers will simply be residents and local business people. How are you going to get them, first, to read this booklet and, second, to use some of its suggestions?

Take several minutes to think about this problem, then write an opening paragraph for the booklet that includes a shared goal.

To test the effectiveness of your paragraph, consider the following two points about shared goals:

1. A shared goal can *motivate your audience* to read and remember what you have to say. Does your paragraph suggest that the booklet will solve some problem your reader faces or achieve some end he or she really cares about? An appeal to vague goals or a wishy-washy generalization such as "our heritage" is unlikely to keep the reader interested. Use your knowledge to fill some need your reader really has.

In a professional situation think of it this way: Your reader has ten letters and five reports on her desk this morning. Your opening statement with its shared goal should tell her why she would want to read your report first and read it carefully.

2. A shared goal can *increase comprehension.* People understand and retain information best when they can fit it into a framework they already know. For example, the context of "home repair" and "do-it-yourself" would be familiar and maybe attractive to your readers. In contrast, if you defined the goal as "architectural renovation" or "techniques of historical landmark preservation" you would make sense to members of the historical society but would have missed your primary audience, the local readers. They would probably find that context not only unfamiliar but somewhat intimidating.

Offer your readers a shared goal—one for which they already have a framework—that helps them turn your message into something meaningful to them.

Here are examples of three different introductory paragraphs written for this booklet. After reading each one, consider how you would evaluate its power to motivate and aid comprehension. Then read the reaction of another reader, which follows each paragraph.

1. This booklet will help you create civic pride and preserve our city's heritage. In addition you will be helping the Historical Society to grow and extend its influence over the city.

A reader's response:

I suppose civic pride is a good thing, but I'm not sure I'd want to help create it. This paragraph makes me feel a little suspicious. What does the Historical Society want from me? I'll bet this is going to be a booklet about raising money so they can put up city monuments.

2. This booklet is concerned with civic restoration and maintenance projects in designated historical areas. It discusses the methods and materials approved by the City Historical Society and City Board of Engineers.

A reader's response:

I guess this is some booklet for city planners or the people who want to set up museums. "Methods and materials" must refer to all those rules and regulations that city contractors have to follow. I wouldn't want to get mixed up with all that if I were doing improvements on my own home.

3. If you own an older home or historical building, there are a number of ways you can preserve its beauty and historical value. At the same time you can increase its market value and decrease its maintenance costs. This booklet will show you five major ways to improve your building and give you step-by-step procedures for how to do this. Please read the booklet over and see which of the suggestions might be useful to you.

A reader's response:

This might be a good idea. I don't know if I'd want to buy the whole package, but I think I'll read it over. What is it—five things I could do? I might find something useful I could try out. I'm particularly concerned about the maintenance costs. Maybe I can find something here on insulation.

Note that in the final example the writer not only has identified shared goals but has given the reader a sort of mental map for reading and understanding the rest of the booklet.

Sometimes a shared goal is something as intangible as intellectual curiosity. But it is the writer's job, in whatever field, to recognize goals or needs that his reader might have and to try to fulfill them. Philosopher Bertrand Russell set forth his shared goal in this way in his introduction to *A History of Western Philosophy:*

Why, then, you may ask, waste time on such insoluble problems? To this one may answer as a historian, or as an individual facing the terror of cosmic loneliness. . . . to teach how to live without certainty, and yet without being paralyzed by hesitation, is perhaps the chief thing that philosophy, in our age, can still do for those who study it.*

To sum up, the first step in designing your paper for a reader is to set up a shared goal. Use it in your problem/purpose statement, and you might also use it as the top level of your issue tree when you generate ideas. A good shared goal will motivate your reader by providing a context for understanding your ideas and a reason for acting on them.

* Bertrand Russell. *A History of Western Philosophy* (New York: Simon and Schuster, 1945), p. xiv.

Strategy 2: Develop a reader-based structure

Most of us intend to write reader-based prose, to communicate with our reader. But for various reasons, people often end up writing writer-based prose, or talking to themselves. For example, the following excerpts are from letters written by students applying for summer jobs. They had been asked to include some personal background and experience.

Do some detective work on these paragraphs and try to describe the hidden logic that you think is organizing each one. Compare the paragraph to some other way you could write it. Why did the writers choose to include the particular facts they did, and why did they organize them in these particular ways?

Terry F.:

I was born in Wichita, Kansas, on December 4, 1962. After four years there my family moved to Topeka, Kansas, where I attended kindergarten. The next year my family moved to Rose Hill, Iowa. I went to first grade there and my family moved again. I started second grade in Butler, Pennsylvania, and finished it in Pittsburgh, Pennsylvania, where I still live today. . . . I took the college curriculum in high school, which included English, history, science, French, and mathematics, and am currently a college sophomore.

I would like this job for two reasons. First, I could use the money for school next year. Second, the experience would be very helpful. It would help me get a job in that specific area when I graduate.

Katherine P.:

As a freshman I worked as a clerk in a student-managed store, Argus. . . . I became acquainted with the university personnel manager and was offered the position of Argus personnel manager for the semester beginning August, 1979. I accepted and held the job until December of 1979, when a managerial position was eliminated. With managerial staff reduced to two people, responsibilities were adjusted and I was offered the position of purchasing agent. Again I accepted.

Notice that in these examples there is a logic organizing each paragraph, but it is the logic of a story, based on the writer's own memories and, in Terry's case, personal needs. The needs of the reader have not been considered. This is writer-based prose: writing that may seem quite clear and organized to the writer but is not yet adequately designed for the reader. In each case, the potential employer probably wanted to know how the applicant's background and experience could fit his or her needs. But neither paragraph was organized around that goal.

Why do people write to themselves when they are ostensibly writing to a reader? One reason is a natural mental habit that psychologists call "egocentrism": thinking centered around the ego or "I." Egocentrism is not selfishness but simply the failure to actively imagine the point of view of someone else as we talk or write. We see this all the time in young children who happily talk about what they are doing in a long, spirited monologue that has many gaps and mysterious expressions. They may speak in code words or private language which, like jargon in adults, is saturated with meaning for the user but not for the listener. Although a bystander may be totally in the dark, the child seems to assume everyone understands perfectly.

Part of a child's cognitive development is growing out of this self-centeredness and learning to imagine and adapt to another person's state of mind. But we never grow out of our egocentrism entirely. When adults write to themselves, it is usually because they have simply forgotten to consider the reader.

There is another very good reason adults write writer-based prose. If you are working on a difficult paper, it is often easier to discover what you know first and worry about designing it for a reader later. An interesting study called the New York Apartment Tour experiment* demonstrated people's tendency to explain in a self-oriented way. The experimenters, Charlotte Linde and William Labov, posed as social workers and asked a number of people to describe their apartments. They found that nearly everyone gave them a room-by-room verbal tour and used similar procedures for conducting it. Although neither the experimenters nor the speakers were actually in the apartments, the descriptions were phrased as though they had been. For example, the description typically starts at the door; if the nearest room is a big one, you go on in ("from the left of the hall you go into the living room"); if the nearest room is small, the speaker merely refers to it and makes a comment ("and there's a closet off the living room"). Then the speaker suddenly brings you back to the entrance hall ("and on the right of the hall is the dining room"), without having to retrace steps or repeat previous rooms. The intuitive, narrative procedures used in conducting this verbal tour were very efficient for remembering all the details of the apartment.

Linde and Labov found that 97 percent of the people questioned used this sort of *narrative* tour strategy. Only 3 percent gave an *overview* such as, "Well, the apartment is basically a square." The reason? The narrative tour strategy is a very efficient way to retrieve information from memory—that is, to survey what you know. In this case it allows you to cover all the rooms one by one as you walk through your apartment. Yet it is almost impossible for another person to reproduce the apartment from this narrative tour, whereas the overview approach, which only 3 percent used, works quite well. As in writing, an organization that functions well for thinking about a topic often fails to communicate that thinking to the listener. A strategy that is effective for the speaker may be terribly confusing to a listener.

---

* Charlotte Linde and William Labov, "Spatial Networks as a Site for the Study of Language and Thought," *Language*, 51 (1975), pp. 924–39.

Note, however, that in draft form, writer-based prose can have a real use. Since this type of writing comes naturally to us, it can be an efficient strategy for exploring a topic and outwitting our nemesis, short-term memory. If a writer's material is complicated or confusing, he may initially have to concentrate all his attention on generating and organizing his own knowledge. He might simply be too preoccupied to simultaneously imagine another person's point of view and adapt to it. The reader has to wait. But you don't want to make the reader wait forever.

You can usually recognize writer-based prose by one or more of these features:

1. An *egocentric focus* on the writer.

2. A *narrative organization* focused on the writer's own discovery process.

3. A *survey structure* organized, like a textbook, around the writer's information.

There are times, of course, when a narrative structure is exactly right—if, for example, your goal is to tell a story or describe an event. And a survey of what you know can be a reasonable way to organize a background report or survey. But in most expository and persuasive writing, the writer needs to reorganize his or her knowledge around a problem, a thesis, or the reader's needs. Writer-based prose just hasn't been reorganized yet.

A reader's test

Below are two versions of a report that will be used as a test case. The writers were students in an organizational psychology course who were also working as consultants to a local organization, the Oskaloosa Brewing Company. The purposes of the report were to show progress to their professor and to present a problem analysis, complete with causes and conclusions, to their client. Both readers—academic and professional—were less concerned with what the students had done or seen than with *how* they had approached the problem and *what* they had made of their observations.

To gauge the reader-based effectiveness of this report, read quickly through Draft 1 and imagine the response of Professor Charns, who needed to answer these questions: "As analysts, what assumptions and decisions did my students make? Why did they make them? And at what stage in the project are they now?" Then reread the draft and play the role of the client, who wants to know: "How did they define the problem, and what did they conclude?" As either reader, can you quickly extract the information the report should be giving you? Next try the same test on Draft 2.

Draft 1

Group Report

(1)   Work began on our project with the initial group decision to evaluate the Oskaloosa Brewing Company. Oskaloosa Brewing Company is a regionally located brewery manufacturing several different types of beer, notably River City and Brough Cream Ale. This beer is marketed under various names in Pennsylvania and other neighboring states. As a group, we decided to analyze this organization because two of our group members had had frequent customer contact with the sales department. Also, we were aware that Oskaloosa Brewing had been losing money for the past five years, and we felt we might be able to find some obvious problems in its organizational structure.

(2)   Our first meeting, held February 17th, was with the head of the sales department, Jim Tucker. Generally, he gave us an outline of the organization, from president to worker, and discussed the various departments that we might ultimately decide to analyze. The two that seemed the most promising and more applicable to the project were the sales and production departments. After a few group meetings and discussions with the personnel manager, Susan Harris, and our advisor, Professor Charns, we felt it best suited our needs and Oskaloosa Brewing's needs to evaluate their bottling department.

(3)   During the next week we had a discussion with the superintendent of production, Henry Holt, and made plans for interviewing the supervisors and line workers. Also, we had a tour of the bottling department that gave us a first-hand look at the production process. Before beginning our interviewing, our group met several times to formulate appropriate questions to use in interviewing, for both the supervisors and the workers. We also had a meeting with Professor Charns to discuss this matter.

(4)   The next step was the actual interviewing process. During the weeks of March 14-18 and March 21-25, our group met several times at Oskaloosa Brewing and interviewed ten supervisors and twelve workers. Finally, during this past week, we have had several group meetings to discuss our findings and the potential problem areas within the bottling department. Also, we have spent time organizing the writing of our progress report.

(5)   The bottling and packaging division is located in a separate building, adjacent to the brewery, where the beer is actually manufactured. From the brewery the beer is piped into one of five lines (four bottling lines and one canning line) in the bottling house, where the bottles are filled, crowned, pasteurized, labeled, packaged in cases, and either shipped out or stored in the warehouse. The head of this operation, and others, is production manager Phil Smith. Next in line under him in direct control of the bottling house is the superintendent of bottling and packaging, Henry Holt. In addition, there are a total of ten supervisors who report directly to Henry Holt and who oversee the daily operations and coordinate and direct the twenty to thirty union workers who operate the lines.

(6)   During production, each supervisor fills out a data sheet to explain what was actually produced during each hour. This form also includes the exact time when a breakdown occurred, what it was caused by, and when production was resumed. Some supervisors' positions are production-staff-oriented. One takes care of supplying the raw material (bottles, caps, labels, and boxes) for production. Another is responsible for the union workers' assignments each day.

These workers are not all permanently assigned to a production-line position. Men called ''floaters'' are used, filling in for a sick worker or helping out after a breakdown.

(7)   The union employees are generally older than thirty-five, some in their late fifties. Most have been with the company many years and are accustomed to having more workers per a slower moving line. . . .

Draft 2

MEMORANDUM

TO: Professor Martin Charns

FROM: Nancy Lowenberg, Todd Scott, Rosemary Nisson,
       Larry Vollen

DATE: March 31, 1977

RE: Progress Report: The Oskaloosa Brewing Company

Why Oskaloosa Brewing?

   Oskaloosa Brewing Company is a regionally located brewery manufacturing several different types of beer, notably River City and Brough Cream Ale. As a group, we decided to analyze this organization because two of our group members have frequent contact with the sales department. Also, we were aware that Oskaloosa Brewing had been losing money for the past five years and we felt we might be able to find some obvious problems in its organizational structure.

Initial Steps: Where to Concentrate?

   After several interviews with top management and a group discussion, we felt it best suited our needs, and Oskaloosa Brewing's needs, to evaluate the production department. Our first meeting, held February 17, was with the head of the sales department, Jim Tucker. He gave us an outline of the organization and described the two major departments, sales and production. He indicated that there were more obvious problems in the production department, a belief also suggested by Susan Harris, the personnel manager.

Next Step

   The next step involved a familiarization with the plant and its employees. First, we toured the plant to gain an understanding of the brewing and bottling processes. Next, during the weeks of March 14-18 and March 21-25, we interviewed ten supervisors and twelve workers. Finally, during the past week we had group meetings to exchange information and discuss potential problems.

The Production Process
    Knowledge of the actual production process is impera-
tive in understanding the effects of various problems on
efficient production. Therefore, we have included a brief
summary of this process.
    The bottling and packaging division is located in a
separate building, adjacent to the brewery, where the beer
is actually manufactured. From the brewery the beer is
piped into one of five lines (four bottling lines and one
canning line) in the bottling house, where the bottles are
filled, crowned, pasteurized, labeled, packaged in cases,
and either shipped out or stored in the warehouse.

Problems
    Through extensive interviews with supervisors and
union employees, we have recognized four apparent problems within the bottling house operations. The first is
that the employees' goals do not match those of the company. . . . This is especially apparent in the union employees, whose loyalty lies with the union instead of the
company. This attitude is well-founded, as the union en-
sures them of job security and benefits. . . .

As a reader, how would you describe the difference between these two versions? Each was written by the same group of writers, but the revision came after a discussion about what the readers really needed to know and expected to get from the report. Let us look at the three things that make Draft 1 a piece of writer-based prose.

Narrative organization

The first four paragraphs of the draft are organized as a narrative, starting with the phrase, "Work began. . . ." We are given a story of the writers' discovery process. Notice how all of the facts are presented in terms of *when* they were discovered, not in terms of their implications or logical connections. The writers want to tell us what happened when; the reader, on the other hand, wants to ask "why?" and "so what?"

A narrative organization is tempting to write because it is a prefabricated order and easy to generate. Instead of having to create a hierarchical organization among ideas or worry about a reader, the writer can simply remember his or her own discovery process and write a story. Papers that start out, "In studying the economic causes of World War I, the first thing we have to consider is. . . ." are often a dead giveaway. They tell us we are going to watch the writer's mind at work and follow him through the process of thinking out his conclusions.

This pattern has, of course, the virtue of any form of drama—it keeps you in suspense by withholding closure. But only if the audience is willing to wait that long for the point. Unfortunately, most academic and professional readers are impatient and tend to interpret such narrative, step-by-step structures either as wandering and confused (does he have a point?) or as a form of hedging.

Egocentric focus

The second feature of Draft 1 is that it is a discovery story starring the writers. Its drama, such as it is, is squarely focused on the writer: "I did/I thought/I felt." Of the fourteen sentences in the first three paragraphs, ten are grammatically focused on the writers' thoughts and actions rather than on the issues. For example: "Work began . . . ," "We decided . . . ," "Also we were aware . . . and we felt. . . ." Generally speaking, the reader is more interested in issues and ideas than in the fact that the writer thought them.

Survey form or textbook organization

In the fifth paragraph of Draft 1, the writers begin to organize their material in a new way. Instead of a narrative, we are given a survey of what the writers observed. Here, the raw facts of the bottling process dictated the organization of the paragraph. Yet the client-reader already knows this, and the professor probably doesn't care. In the language of computer science we could say the writers are performing a "memory dump": simply printing out information in the exact form in which they stored it in memory. Notice how in the revised version the writers try to *use* their observations to understand production problems.

The problem with a survey or "textbook" form is that it ignores the reader's need for a different organization of the information. Suppose, for example, you are writing to model airplane builders about wind resistance. The information you need comes out of a physics text, but that text is organized around the field of physics; it starts with subatomic particles and works up from there. To meet the needs of your reader, you have to adapt that knowledge, not lift it intact from the text. Sometimes writers can simply survey their knowledge, but generally the writer's main task is to *use* knowledge rather than reprint it.

To sum up, in Draft 2 of the Oskaloosa report, the writers made a real attempt to write for their readers. Among other things the report is now organized around major questions readers might have, it uses headings to display the overall organization of the report, and it makes better use of topic sentences that tell the reader what each paragraph contains and why to read it. Most important, it focuses more on the crucial information the reader wants to obtain.

Obviously this version could still be improved. But it shows the writers attempting to transform writer-based prose and change their narrative and survey pattern into a more issue-centered hierarchical organization.

Consider another example of how a writer transformed a writer-based paragraph into a reader-based one. The first draft below is full of good ideas but has a narrative organization and egocentric focus. We can almost see the writer reading the book. Her conclusions (which her professor will want to know) are buried within a description of the story (which her professor, of course, knows already).

Writer-based draft:

In *Great Expectations,* Pip is introduced as a very likable young boy. Although he steals, he does it because he is both innocent and good-hearted. Later, when he goes to London, one no longer feels this same sort of identification with Pip. He becomes too proud to associate with his old friends, cutting ties with Joe and Biddy because of his false pride. And yet one is made to feel that Pip is still an innocent in some important ways. When he dreams about Estella, one can see how all his unrealistic, romantic illusions blind him to the way the world really works.

We know from this paragraph how the writer reacted to a number of things in the novel. But what conclusions did she finally come to? What larger pattern does she want us to see?

Reader-based revision:

In *Great Expectations* Pip changes from a goodhearted boy into a selfish young man, yet he always remains an innocent who never really understands how the world works. Although as a child Pip actually steals something, he does it because he has a gullible, kindhearted sort of innocence. As a young man in London his crime seems worse when he cuts his old friends, Joe and Biddy, because of false pride. And yet, as his dreams about Estella show, Pip is still an innocent, a person caught up in unrealistic romantic illusions that he can't see through.

The revised version starts out with a topic sentence that explicitly states the writer's main idea and shows us how she has chunked or organized the facts of the novel. The rest of the paragraph is clearly focused on that idea, and words such as "although" are used to show how her observations are logically related to one another. From a professor or other reader's point of view, this organization is also more effective because it clearly shows what the writer learned from reading the novel.

Below is a good example of a writer who has focused all of his attention on the object before him. He has given us a survey of what he knows about running shoes, although the ostensible purpose of the paragraph was to help a new runner decide what shoe to buy.

Writer-based draft:

Shoes are the most important part of your equipment, so choose them well. First, there are various kinds. Track shoes are lightweight with spikes. Road running flats, however, are sturdy, with ½″ to 1″ of cushioning. In many shoes the soles are built up with different layers of material. The uppers are made in various ways, some out of leather, some out of nylon reinforced with leather, and the cheapest are made of vinyl. The best combination is nylon with a leather heel cup. The most distinctive thing about running shoes is the raised heel and, of course, the stripes. Although some tennis shoes now have such stripes, it is important not to confuse them with a real running shoe. All in all, a good running shoe should combine firm foot support with sufficient flexibility.

In this draft the writer has focused on the shoe, not the reader who needs to choose a shoe. How would we decide between leather, nylon, and vinyl? Or judge what is "sufficiently" flexible? Why does it matter that the soles are layered; was the writer trying to make a point?

Reader-based revision:

Your running shoe will be your most important piece of running equipment, so look for a shoe that both cushions and supports your foot. Track shoes, which are lightweight and flimsy, with spikes for traction in dirt, won't do. Neither will tennis shoes, which are made for balance and quick stops, not steady pounding down the road. A good pair of shoes starts with a thick layered sole, at least ½″ to 1″ thick. The outer layer absorbs road shock; the inner layer cushions your foot. Another form of cushioning is the slightly elevated heel which prevents strain on the vulnerable Achilles tendon.

The uppers that will support your foot come in vinyl, which is cheaper but can cause blisters and hot feet; in leather, which can crack with age; and in a lightweight but more expensive nylon and leather combination. The best nylon and leather shoes will have a thick, fitted leather heel cup that keeps your foot from rolling and prevents twisted ankles. Make sure, however, that your sturdy shoes are still flexible enough that you can bend 90° at the ball of your foot. Although most running shoes have stripes, not all shoes with stripes can give you the cushioning and flexible support you need when you run.

Notice how the revision uses the same facts about shoes but organizes them around the reader's probable questions. The writer tells us what his facts *mean* in the context of choosing shoes. For example, vinyl uppers mean low cost and possible blisters. And the topic sentence sets up the key features of a good shoe—cushioning and support—which the rest of the paragraph will develop. The reader-based revision tells us what we need to know in a direct, explicit way.

Creating reader-based prose

In the best of all possible worlds we would all write reader-based prose from the beginning. It is theoretically much more efficient to generate and organize your ideas in light of the reader in the first place. But sometimes that is hard to do. Take the assignment: "Write about the physics of wind resistance for a model airplane builder." For a physics teacher this would be a trivial problem. But for someone ten years out of Physics 101, the first task would be remembering whatever they knew about wind resistance or friction at all. Adapting that knowledge to the reader would just have to wait.

In general, write for your reader whenever you can, but recognize that many times a first draft is going to be more writer-based than you may want it to be. Even though the draft may not work well for your reader, it can represent a great deal of work for you and be the groundwork for an effective paper. The more complex your problem and the more difficult your material, the more you will need to transform your writer-based prose to reader-based prose. This is not an overly difficult step in the writing process, but many writers simply neglect to take it.

In order to transform your paper to more reader-based prose, there are four major things you can do, all of which should be familiar by now:

1. Organize your paper around a problem, a thesis, or a purpose you share with the reader—not around your own discovery process or the topic itself.

2. With a goal or thesis as the top level of your issue tree, organize your ideas in a hierarchy. Distinguish between your major and minor ideas and make the relationship between them explicit to the reader. You can use this technique to organize not only an entire paper but sections and paragraphs.

3. If you are hoping that your reader will draw certain conclusions from your paper, or even from a portion of it, make those conclusions explicit. If you expect him or her to go away with a few main ideas, don't leave the work of drawing inferences and forming concepts up to your reader. He or she might just draw a different set of conclusions.

4. Finally, once you have created concepts and organized your ideas in a hierarchy focused on your reader and your goals, use cues—which we will discuss shortly—to make that organization vivid and clear to the reader.

Strategy 3: Give your reader cues

Part of your contract with a reader, if you seriously want to communicate, is to guide him or her through your prose. You need to set up cues that help the reader see what is coming and how it will be organized. This means, first of all, creating expectations and fulfilling them so that when your point arrives, your reader will have a well-anchored hook to hang it on. . . . In addition, you want the reader to know which points are major, which are minor, and how they are related to one another. By using various kinds of cues and signposts, you can guide the reader to build an accurate mental tree of your discussion.

Readers, of course, come to your prose with built-in expectations about where these cues will be. For example, they expect:

- To find the most important points of a discussion stated at the beginning and summarized in some way at the end.
- To find a topic sentence that tells them what they will learn from a paragraph.
- To find the writer's key words in grammatically important places such as the subject, verb, and object positions.

It is to your advantage to fulfill these expectations whenever you can.

Writers have a number of tools and techniques they can use to *preview* their meaning, *summarize* it, and *guide* the reader. Figure 1 lists some of the most common. Check this list against the last paper you wrote. How many of these tools did you take advantage of?

Fig. 1. Cues for the Reader

Title Table of contents Abstracts Introduction Headings Problem/purpose statement Topic sentences for paragraphs	*Cues that preview your points*
Sentence summaries at ends of paragraphs Conclusion or summary sections	*Cues that summarize or illustrate your points*

Pictures, graphs, and tables
Punctuation
Typographical cues:
    different typefaces,
    underlining, numbering
Visual arrangement:
    indentation, extra white space,
    rows and columns
	Cues that guide the reader visually

Transitional words Conjunctions Repetitions Pronouns Summary nouns	*Cues that guide the reader verbally*

The conventions of format on a page also work as familiar cues to the reader. Figure 2 shows a typical format for papers and reports.

Draft 2 of the Oskaloosa Brewing report . . . offered a good example of
how headings, topic sentences, and previews of conclusions can provide reader
cues. Here is another piece of writing that was designed with the reader in
mind. It comes from Thomas Miller's book *This Is Photography,** in a chapter
called "Action." One of the first previews the reader sees on the page is a
photo of a pole vaulter effortlessly sailing over a bar and a place kicker completely off the ground with his right foot at the top of his kick. The caption
reads, "These look like top speed but. . . ."

Fig. 2. Common Format for Typewritten Paper

                    THIS IS A TITLE: THE SUBTITLE QUALIFIES IT

        The first sentence in this paragraph is a topic sentence,
    which announces the topic and previews the argument or point
    of the paragraph. The remainder of the paragraph often pre-
    views the rest of the paper, introducing the main points to
    be covered.

    THIS IS A MAJOR HEADING

        Major headings are placed flush left, often set in caps,
    and, in typewritten manuscript, usually underlined. In print
    they are often set in boldface type. Ideally a reader should be
    able to see the shape of your discussion simply by reading the
    title and major headings. Make the wording of major headings
    grammatically parallel, if you can, as the major and minor
    headings are in this example.

        This Is a Minor Heading

        Unlike a major heading it is indented and typed in capital
    and lower-case letters. It should be clearly and logically re-
    lated to the major heading that precedes it.

            The fact that this passage is indented says it
            is either a long quotation or an example. The
            additional space around it and the single spac-
            ing signal that it is a different kind of text,
            and let readers adjust their reading speed and
            expectations.

---

* Thomas Miller and Wyatt Brummitt, *This Is Photography* (Rochester, N.Y.: Case Hoyt Corp.,
  1945).

In the passage below, I have italicized and footnoted certain portions for discussion later. As you read the italicized parts, try to figure out what effect the writer was hoping to have on you by using the cues he did.

Poised Action[1]

In many sports,[2] particularly in races, movement is constant enough to permit picture making in terms of calculated speeds. *But there are other sports*[2] in which the action is spasmodic, defying calculation. *In those sports,*[2] the instants when action is poised are, pictorially, just as vivid and interesting as the moments when action is wildest. *Take pole vaulting, for instance.*[2] At the very top of the vault, with the vaulter's body flung out horizontally over the bar, action is relatively quiet—yet it's the best pictorial moment in this field event. This peak instant can be "stopped" with much less shutter speed than either the rise or fall.

Baseball[3] has a number of moments which are full of *poised action.*[3] The pitcher winds up and *then*[4] unwinds to throw his speed-ball. *In that instant,*[4] between winding and unwinding, action is suspended, yet a picture of it tells a story of speed and power. *An instant later,*[4] having released the ball, the pitcher is *again*[4] poised—all his energy having gone into the delivery. *There's another pictorial moment.*[5] *To picture either of these moments you need to work swiftly, but a high shutter speed is less important to your success than an understanding of the sport and of the personal style of the athlete before your lens.*[6]

Even in boxing,[7] a good photographer gets his pictures as the blows land, not as they travel. *There was that famous instance*[8] at the Louis-Nova fight in '41. Two photographers, on directly opposite sides of the ring, saw a heavy punch coming and shot just as it landed. Both used Photoflashes, of course, but—one of the lamps failed to work. The photographer whose light had failed discovered, on developing the film, that he had a picture—a most unusual and vivid silhouette—*made by the light of his competitor's flash.*[9] The fighters hid the other man's flash bulb, so the silhouette effect was perfect—and dramatic. *The only moral to this yarn*[10] is that experience teaches pressmen and other pro's that there are right instances for any shots. The photographers on opposite sides of the ring were right—and right together, within the same hundredth of a second.

Here are comments on the writer's cues:

    1. In the original, this heading is set in boldface type.

    2. These cues make the relationship between each of the sentences explicit. They lead us along; many sports are contrasted to other sports. We are told something additional about "other sports" and then given an example.

    3. A topic sentence ties a new subject, "baseball," to the old topic, "poised action."

4. These words and phrases reinforce our sense of the timing and sequence of the action.

5. The writer recaps his discussion by redefining it not just as an action, but in the larger context now of photographs representing "pictorial moments."

6. This sentence is a recap on an even larger scale. In it the writer draws a conclusion based on both this paragraph and the preceding one, and ties the paragraphs to the larger goal of the book and the chapter: how to take good action photographs.

7. This topic sentence and its introductory phrase are performing two functions: they introduce a new subject, boxing, and tie it neatly to the old framework with the words "even in."

8. We are told to see this as an example of the writer's point. He doesn't let us simply be entertained by the story; he uses it.

9. This line was also in italics in the original, to emphasize how unusual the occurrence was. Note that in the phrase just before this one— "a most unusual and vivid silhouette"—the writer used dashes to highlight the significance of the facts. Both italics and dashes are attention-getting cues, though they can be overdone.

10. The writer draws a particular conclusion from all of this that is tied to the point of his book, and he signposts his conclusion quite clearly so we won't miss it: "The only moral to this yarn is . . ."

Strategy 4: Develop a persuasive argument

People often write because they want to make something happen: they want the reader to do something or at least to see things their way. But sometimes expressing a point of view isn't enough, because it conflicts with the way the reader *already* sees things. We are faced with the same old problem of communication: your image of something and your reader's are not the same. What kinds of argument can you use that will make him or her see things *your* way? In this section we will look at the nature of arguments and at one type, the Rogerian argument, that can help you persuade another person to see things differently.

Winning an argument versus persuading a listener

When people think of arguments they usually think of winning them. And the time-honored method of winning an argument is by force ("You agree or I'll shoot.") or, in its more familiar form, by authority ("This is right because I [your mother, father, teacher, sergeant, boss] say it is."). The problem with force or authority is that, short of brainwashing, it often changes people's behavior but not their minds.

A second familiar form of argument is debate. Yet many people who learn to debate in high school discover that in the real world their debate strategies can indeed prove their point—but lose the argument. Debate is an argumentative contest: person A is pitted against person B, and the winner is decided by an impartial judge. But in the real world, person A is trying not to impress a judge but to *convince* person B. The goal of such an argument is not to win points but to affect your listener, to change his or her image of your subject to some significant way. And, as you remember, that image may be a large, complex network of ideas, associations, and attitudes. The goal of communication is to find a common ground and create a shared image, but debate typically polarizes a discussion by putting one image against the other.

Let us look for a moment at the possible outcomes of an argument or discussion in which the two parties have firmly held but differing images. Ann has decided to take a year off to work and travel before she finishes college and settles on a career. Her parents immediately oppose the idea. To them, this plan conveys an image of "dropping out" and wasting a year, with the possibility that Ann might not return to school. Furthermore, they have saved money to help put her through school and see this prospect as an indication that she doesn't value their plans, hopes, and efforts for her.

For Ann, on the other hand, taking a year off means getting time and experience that would enable her to take better advantage of college. She hopes it will help her decide what sort of work she wants to do, but more importantly she sees it as a chance to develop on her own for a while. In her mind, the goal of going to college isn't getting a degree but figuring out what things you want to learn more about.

Clearly Ann and her parents have very different images of taking a year off. Assume you are Ann in this situation. What are the possible outcomes of an argument you might have with your parents?

One outcome, and usually the least likely one, is that you will totally reconstruct your listeners' image so they see the issue just as you do. You simply replace their perspective with yours. Reconstruction can no doubt happen if your audience has an undeveloped image of the subject or sees you as a great authority, but argument strategies that set out to reconstruct someone else's ideas completely—to *win* the point—are usually ill-founded and unrealistic. They are more likely to polarize people than to persuade them.

A second alternative is to modify someone else's image, to add to or clarify it. You do this when you clarify an issue (for example, taking a year off is not the same as "dropping out") or when you add new information (Ann's college even has a special program for this and might give her some course credit for work experience). As a writer this is clearly the most reasonable effect you can aim for. In doing so you respect the other person's point of view while striving to modify those features you can reasonably affect.

The third possible outcome of an argument may be the most common: no change. Think for a minute of how many speeches, lectures, classes, sermons, and discussions you have sat through in your life and how many of those had no discernible effect on your thinking. If we think of an argument as debate in which a "good" argument inevitably wins, we forget that it is possible for even a "correct" argument to have absolutely no effect on our listener.

To sum up, the goal of an argument is to modify the image of your listener—and that this is not the same as simply presenting your own image. A successful argument is a reader-based act. It considers attitudes and images the reader already holds.

However, a great roadblock stands in the way of modifying a listener's image. Many people perceive any change in their image of things as a threat to their own security and stability. People's images are part of themselves, and a part of how they have made sense of the world. To ask them to change their image in any significant way can make people anxious and resistant to change. When this happens, communication simply stops.

Arguments that polarize issues often create just this situation. The more the speaker argues, the more firmly the listener clings to his own position. And instead of listening, the listener spends his time thinking up counter-arguments to protect his own position and image. So the critical question for the writer is this: how can I persuade my reader to listen to my position and maybe even modify his or her image without creating this sense of threat that stops communication?

Rogerian argument

Rogerian argument, developed in part from the work of Carl Rogers, is an argument strategy designed not to win but to increase communication in both directions. It is based on the fact that if people feel they are understood—that their position is honestly recognized and respected—they may cease to feel a sense of threat. Once the threat is removed, listening is no longer an act of self-defense, and people feel they can afford to truly listen to and consider other ways of seeing things.

The goal of an argument, then, is to induce your reader at least to consider your position and the possibility of modifying his or her own. One way to make this happen is to demonstrate an understanding of your listener's position *first*. That means trying to see the issue from his or her point of view. For face-to-face discussions, Carl Rogers suggested this rule of thumb: before you present your position and argue for your way of seeing things, you must be able to describe your listener's position back to your listener in such a way that he or she *agrees* with your version of it. In other words, you are demonstrating that you not only care about your listener's perspective but care enough to actively try (and keep trying) to understand it. So Ann in our example would have begun the discussion with her parents by exploring with them their response to her leaving college and the reasons behind their feelings.

What does this mean for writers who don't have the luxury of a face-to-face discussion? First, you can use the introduction to your paper, including your shared goal, to demonstrate to your reader a thorough understanding of his or her problems and goals. This is your chance to look at the question from your reader's point of view and show how your message is relevant to them.

Secondly, try to avoid categorizing people and issues. This puts people into camps, polarizes the argument, and stops communication. For example, Ann may well have felt that her parents were being old-fashioned and conventional to resist her idea, but establishing that point would have done little to change their minds. A Rogerian argument, by contrast, would begin by acknowledging the parents' plans and hopes for her and recognizing the element of truth in their fear of her "dropping out." They know that, despite good intentions, many people don't come back to college. In taking a Rogerian approach, Ann might also begin to understand the issues more clearly herself. One of the hidden strengths of a Rogerian argument is that, besides increasing one's power to persuade, it also opens up communication and may even end up persuading the persuader. It increases the possibility of genuine communication and change for both the speaker and listener.

The first draft of Ann's letter started like this:

Dear Mom and Dad,
I wish you would try to see my point of view and not be so conventional. Things are different from when you went to school. And you must realize I am old enough to make my own decisions, even if you disagree. There are a number of good reasons why this is the best decision I could make. First, . . .

Although this letter created a "strong" argument, it was also likely to stop communication and unlikely to persuade. Here is the letter Ann eventually wrote to her parents, which tries to take an open Rogerian approach to the problem.

Dear Mom and Dad,
As I told you the other night on the phone, I want to consider taking a year off from college to work and be on my own for a while. I've been thinking over what you said because this is an important decision and, like you, I want to do what will be best in the long run, not just what seems attractive now. I think some of your objections make a lot of sense. After all the effort you've put into helping me get through college, it would be terrible to just "drop out" or never find a real career that I could be committed to.

I know you're also wondering if I recognize what an opportunity I have and are probably worrying if I'm just going to let it slip through my fingers. Well, in a way I'm worried about that too. Here I am working hard, but I don't really know where I want to go or why. It's time for me to specialize and I can't decide what to do. And it's that opportunity I'm afraid of losing. I feel I need some time off and some experience so I can make a better decision and really take advantage of my last year here.

But there's still the question of whether I would be dropping out. The college actually has a program for people who want to take a year off, and they even encourage you to enter it if you have some idea of what you'd be doing. So, as far as the school is concerned, I'd be in a well-established leave of absence program. But the fact is, people do drop out. They don't always come back. What would a whole year away from school do to me? You're right, I can't really be sure. But I think my reasons are good ones, and I'm working on a plan that would let me earn credit while I work and come back to school with a clearer sense of where I want to go. Can you offer me any more suggestions on ways I could plan ahead?

Love,
ANN

2

Problems with language

Gobbledygook

STUART CHASE

Stuart Chase worked for many years as a
consultant to various government agencies; his
other books include *The Tyranny of Words*
(1938) and *Democracy Under Pressure* (1945).

*S*aid Franklin Roosevelt, in one of his
early presidential speeches: "I see one-third of a nation ill-housed, ill-clad, ill-
nourished." Translated into standard bureaucratic prose his statement would
read:

> It is evident that a substantial number of persons within the Conti-
> nental boundaries of the United States have inadequate financial re-
> sources with which to purchase the products of agricultural communities
> and industrial establishments. It would appear that for a considerable
> segment of the population, possibly as much as 33.3333* of the total,
> there are inadequate housing facilities, and an equally significant propor-
> tion is deprived of the proper types of clothing and nutriment.
>
> * Not carried beyond four places.

This rousing satire on gobbledygook—or talk among the bureaucrats—is
adapted from a report[1] prepared by the Federal Security Agency in an attempt
to break out of the verbal squirrel cage. "Gobbledygook" was coined by an
exasperated Congressman, Maury Maverick of Texas, and means using two, or
three, or ten words in the place of one, or using a five-syllable word where a
single syllable would suffice. Maverick was censuring the forbidding prose of
executive departments in Washington, but the term has now spread to windy
and pretentious language in general.

"Gobbledygook" itself is a good example of the way a language grows.
There was no word for the event before Maverick's invention; one had to say:
"You know, that terrible, involved, polysyllabic language those government
people use down in Washington." Now one word takes the place of a dozen.

A British member of Parliament, A. P. Herbert, also exasperated with
bureaucratic jargon, translated Nelson's immortal phrase, "England expects
every man to do his duty":

> England anticipates that, as regards the current emergency, person-
> nel will face up to the issues, and exercise appropriately the functions
> allocated to their respective occupational groups.

[1] This and succeeding quotations from F.S.A. report by special permission of the author, Milton
Hall.

SHELL PNR

DATE: _____

SALES AGENT'S NAME: _____

TELEPHONE: _____
(Must include Area Code)

AGENCY NAME: _____

STREET NUMBER: _____

CITY/STATE: _____   ZIP CODE: _____

ATC/IATA CODE #: _____

TELETICKETING CODE: _____
(Four Letters)

FORWARD TO: MARVIN BERNSTEIN

A New Zealand official made the following report after surveying a plot of ground for an athletic field:[2]

It is obvious from the difference in elevation with relation to the short depth of the property that the contour is such as to preclude any reasonable developmental potential for active recreation.

Seems the plot was too steep.

An office manager sent this memo to his chief:

Verbal contact with Mr. Blank regarding the attached notification of promotion has elicited the attached representation intimating that he prefers to decline the assignment.

Seems Mr. Blank didn't want the job.

A doctor testified at an English trial that one of the parties was suffering from "circumorbital haematoma."

Seems the party had a black eye.

In August 1952 the U.S. Department of Agriculture put out a pamphlet entitled: "Cultural and Pathogenic Variability in Single-Condial and Hyphaltip Isolates of Hemlin-Thosporium Turcicum Pass."

Seems it was about corn leaf disease.

On reaching the top of the Finsteraarhorn in 1845, M. Dollfus-Ausset, when he got his breath, exclaimed:

The soul communes in the infinite with those icy peaks which seem to have their roots in the bowels of eternity.

Seems he enjoyed the view.

A government department announced:

Voucherable expenditures necessary to provide adequate dental treatment required as adjunct to medical treatment being rendered a pay patient in in-patient status may be incurred as required at the expense of the Public Health Service.

Seems you can charge your dentist bill to the Public Health Service. Or can you?

[2] This item and the next two are from the piece on gobbledygook by W. E. Farbstein, New York *Times*, March 29, 1953.

Legal talk

Gobbledygook not only flourishes in government bureaus but grows wild and lush in the law, the universities, and sometimes among the literati. Mr. Micawber was a master of gobbledygook, which he hoped would improve his fortunes. It is almost always found in offices too big for face-to-face talk. Gobbledygook can be defined as squandering words, packing a message with excess baggage and so introducing semantic "noise." Or it can be scrambling words in a message so that meaning does not come through. The directions on cans, bottles, and packages for putting the contents to use are often a good illustration. Gobbledygook must not be confused with double talk, however, for the intentions of the sender are usually honest.

I offer you a round fruit and say, "Have an orange." Not so an expert in legal phraseology, as parodied by editors of *Labor:*

> I hereby give and convey to you, all and singular, my estate and interests, right, title, claim and advantages of and in said orange, together with all rind, juice, pulp and pits, and all rights and advantages therein . . . anything hereinbefore or hereinafter or in any other deed or deeds, instrument or instruments of whatever nature or kind whatsoever, to the contrary, in any wise, notwithstanding.

The state of Ohio, after five years of work, has redrafted its legal code in modern English, eliminating 4,500 sections and doubtless a blizzard of "whereases" and "hereinafters." Legal terms of necessity must be closely tied to their referents, but the early solons tried to do this the hard way, by adding synonyms. They hoped to trap the physical event in a net of words, but instead they created a mumbo-jumbo beyond the power of the layman, and even many a lawyer, to translate. Legal talk is studded with tautologies, such as "cease and desist," "give and convey," "irrelevant, incompetent, and immaterial." Furthermore, legal jargon is a dead language; it is not spoken and it is not growing. An official of one of the big insurance companies calls their branch of it "bafflegab." Here is a sample from his collection:[3]

> One-half to his mother, if living, if not to his father, and one-half to his mother-in-law, if living, if not to his mother, if living, if not to his father. Thereafter payment is to be made in a single sum to his brothers. On the one-half payable to his mother, if living, if not to his father, he does not bring in his mother-in-law as the next payee to receive, although on the one-half to his mother-in-law, he does bring in the mother or father.

You apply for an insurance policy, pass the tests, and instead of a straightforward "here is your policy," you receive something like this:

[3] Interview with Clifford B. Reeves by Sylvia F. Porter, New York *Evening Post,* March 14, 1952.

This policy is issued in consideration of the application therefor, copy of which application is attached hereto and made part hereof, and of the payment for said insurance on the life of the above-named insured.

Academic talk

The pedagogues may be less repetitious than the lawyers, but many use even longer words. It is a symbol of their calling to prefer Greek and Latin derivatives to Anglo-Saxon. Thus instead of saying: "I like short clear words," many a professor would think it more seemly to say: "I prefer an abbreviated phraseology, distinguished for its lucidity." Your professor is sometimes right, the longer word may carry the meaning better—but not because it is long. Allen Upward in his book *The New Word* warmly advocates Anglo-Saxon English as against what he calls "Mediterranean" English, with its polysyllables built up like a skyscraper.

Professional pedagogy, still alternating between the Middle Ages and modern science, can produce what Henshaw Ward once called the most repellent prose known to man. It takes an iron will to read as much as a page of it. Here is a sample of what is known in some quarters as "pedageese":

Realization has grown that the curriculum or the experiences of learners change and improve only as those who are most directly involved examine their goals, improve their understandings and increase their skill in performing the tasks necessary to reach newly defined goals. This places the focus upon teacher, lay citizen and learner as partners in curricular improvement and as the individuals who must change, if there is to be curriculum change.

I think there is an idea concealed here somewhere. I think it means: "If we are going to change the curriculum, teacher, parent, and student must all help." The reader is invited to get out his semantic decoder and check on my translation. Observe there is no technical language in this gem of pedageese, beyond possibly the word "curriculum." It is just a simple idea heavily ververbalized.

In another kind of academic talk the author may display his learning to conceal a lack of ideas. A bright instructor, for instance, in need of prestige may select a common sense proposition for the subject of a learned monograph—say, "Modern cities are hard to live in" and adorn it with imposing polysyllables: "Urban existence in the perpendicular declivities of megalopolis . . ." et cetera. He coins some new terms to transfix the reader—"mega-decibel" or "strato-cosmopolis"—and works them vigorously. He is careful to add a page or two of differential equations to show the "scatter." And then he

publishes, with 147 footnotes and a bibliography to knock your eye out. If the
authorities are dozing, it can be worth an associate professorship.

While we are on the campus, however, we must not forget that the technical language of the natural sciences and some terms in the social sciences,
forbidding as they may sound to the layman, are quite necessary. Without
them, specialists could not communicate what they find. Trouble arises when
experts expect the uninitiated to understand the words; when they tell the jury,
for instance, that the defendant is suffering from "circumorbital haematoma."

Here are two authentic quotations. Which was written by a distinguished
modern author, and which by a patient in a mental hospital? You will find the
answer at the end of the chapter.

> 1. Have just been to supper. Did not knowing what the woodchuck
> sent me here. How when the blue blue blue on the said anyone can do it
> that tries. Such is the presidential candidate.
> 2. No history of a family to close with those and close. Never shall
> he be alone to be alone to be alone to be alone to be alone to lend a hand
> and leave it left and wasted.

Reducing the gobble

As government and business offices grow larger, the need for doing something
about gobbledygook increases. Fortunately the biggest office in the world is
working hard to reduce it. The Federal Security Agency in Washington,[4] with
nearly 100 million clients on its books, began analyzing its communication
lines some years ago, with gratifying results. Surveys find trouble in three main
areas: correspondence with clients about their social security problems, office
memos, official reports.

Clarity and brevity, as well as common humanity, are urgently needed in
this vast establishment which deals with disability, old age, and unemployment. The surveys found instead many cases of long-windedness, foggy meanings, clichés, and singsong phrases, and gross neglect of the reader's point of
view. Rather than talking to a real person, the writer was talking to himself.
"We often write like a man walking on stilts."

Here is a typical case of long-windedness:

> *Gobbledygook as found:* "We are wondering if sufficient time has
> passed so that you are in a position to indicate whether favorable action
> may now be taken on our recommendation for the reclassification of
> Mrs. Blank, junior clerk-stenographer, CAF 2, to assistant clerk-stenog-
> rapher, CAF 3?"
> *Suggested improvement:* "Have you yet been able to act on our
> recommendation to reclassify Mrs. Blank?"

[4] Now the Department of Health, Education, and Welfare.

Another case:

> Although the Central Efficiency Rating Committee recognizes that there are many desirable changes that could be made in the present efficiency rating system in order to make it more realistic and more workable than it now is, this committee is of the opinion that no further change should be made in the present system during the current year. Because of conditions prevailing throughout the country and the resultant turnover in personnel, and difficulty in administering the Federal programs, further mechanical improvement in the present rating system would require staff retraining and other administrative expense which would seem best withheld until the official termination of hostilities, and until restoration of regular operations.

The F.S.A. invites us to squeeze the gobbledygook out of this statement. Here is my attempt:

> The Central Efficiency Rating Committee recognizes that desirable changes could be made in the present system. We believe, however, that no change should be attempted until the war is over.

This cuts the statement from 111 to 30 words, about one-quarter of the original, but perhaps the reader can do still better. What of importance have I left out?

Sometimes in a book which I am reading for information—not for literary pleasure—I run a pencil through the surplus words. Often I can cut a section to half its length with an improvement in clarity. Magazines like *The Reader's Digest* have reduced this process to an art. Are long-windedness and obscurity a cultural lag from the days when writing was reserved for priests and cloistered scholars? The more words and the deeper the mystery, the greater their prestige and the firmer the hold on their jobs. And the better the candidate's chance today to have his doctoral thesis accepted.

The F.S.A. surveys found that a great deal of writing was obscure although not necessarily prolix. Here is a letter sent to more than 100,000 inquirers, a classic example of murky prose. To clarify it, one needs to *add* words, not cut them:

> In order to be fully insured, an individual must have earned $50 or more in covered employment for as many quarters of coverage as half the calendar quarters elapsing between 1936 and the quarter in which he reaches age 65 or dies, whichever first occurs.

Probably no one without the technical jargon of the office could translate this: nevertheless, it was sent out to drive clients mad for seven years. One poor fellow wrote back: "I am no longer in covered employment. I have an outside job now."

Many words and phrases in officialese seem to come out automatically, as if from lower centers of the brain. In this standardized prose people never *get jobs,* they "secure employment"; *before* and *after* become "prior to" and "subsequent to"; one does not *do,* one "performs"; nobody *knows* a thing, he is "fully cognizant"; one never *says,* he "indicates." A great favorite at present is "implement."

Some charming boners occur in this talking-in-one's-sleep. For instance:

> The problem of extending coverage to all employees, regardless of size, is not as simple as surface appearances indicate.
>
> Though the proportions of all males and females in ages 16–45 are essentially the same . . .
>
> Dairy cattle, usually and commonly embraced in dairying . . .

In its manual to employees, the F.S.A. suggests the following:

INSTEAD OF	USE
give consideration to	consider
make inquiry regarding	inquire
is of the opinion	believes
comes into conflict with	conflicts
information which is of a confidential nature	confidential information

Professional or office gobbledygook often arises from using the passive rather than the active voice. Instead of looking you in the eye, as it were, and writing "This act requires . . ." the office worker looks out of the window and writes: "It is required by this statute that . . ." When the bureau chief says, "We expect Congress to cut your budget," the message is only too clear; but usually he says, "It is expected that the departmental budget estimates will be reduced by Congress."

GOBBLED: "All letters prepared for the signature of the Administrator will be single spaced."

UNGOBBLED: "Single space all letters for the Administrator." (Thus cutting 13 words to 7.)

Only people can read

The F.S.A. surveys pick up the point . . . that human communication involves a listener as well as a speaker. Only people can read, though a lot of writing seems to be addressed to beings in outer space. To whom are you talking? The sender of the officialese message often forgets the chap on the other end of the line.

A woman with two small children wrote the F.S.A. asking what she should do about payments, as her husband had lost his memory. "If he never gets able to work," she said, "and stays in an institution would I be able to draw any benefits? . . . I don't know how I am going to live and raise my children since he is disable to work. Please give me some information. . . ."

To this human appeal, she received a shattering blast of gobbledygook, beginning, "State unemployment compensation laws do not provide any benefits for sick or disabled individuals . . . in order to qualify an individual must have a certain number of quarters of coverage . . ." et cetera, et cetera. Certainly if the writer had been thinking about the poor woman he would not have dragged in unessential material about old-age insurance. If he had pictured a mother without means to care for her children, he would have told her where she might get help—from the local office which handles aid to dependent children, for instance.

Gobbledygook of this kind would largely evaporate if we thought of our messages as two way—in the above case, if we pictured ourselves talking on the doorstep of a shabby house to a woman with two children tugging at her skirts, who in her distress does not know which way to turn.

Results of the survey

The F.S.A. survey showed that office documents could be cut 20 to 50 per cent, with an improvement in clarity and a great saving to taxpayers in paper and payrolls.

A handbook was prepared and distributed to key officials.[5] They read it, thought about it, and presently began calling section meetings to discuss gobbledygook. More booklets were ordered, and the local output of documents began to improve. A Correspondence Review Section was established as a kind of laboratory to test murky messages. A supervisor could send up samples for analysis and suggestions. The handbook is now used for training new members; and many employees keep it on their desks along with the dictionary. Outside the Bureau some 25,000 copies have been sold (at 20 cents each) to individuals, governments, business firms, all over the world. It is now used officially in the Veterans Administration and in the Department of Agriculture.

The handbook makes clear the enormous amount of gobbledygook which automatically spreads in any large office, together with ways and means to keep it under control. I would guess that at least half of all the words circulating around the bureaus of the world are "irrelevant, incompetent, and immaterial"—to use a favorite legalism; or are just plain "unnecessary"—to ungobble it.

My favorite story of removing the gobble from gobbledygook concerns the Bureau of Standards at Washington. I have told it before but perhaps the reader will forgive the repetition. A New York plumber wrote the Bureau that he had found hydrochloric acid fine for cleaning drains, and was it harmless? Washington replied: "The efficacy of hydrochloric acid is indisputable, but the chlorine residue is incompatible with metallic permanence."

The plumber wrote back that he was mighty glad the Bureau agreed with him. The Bureau replied with a note of alarm: "We cannot assume responsibil-

[5] By Milton Hall.

ity for the production of toxic and noxious residues with hydrochloric acid, and suggest that you use an alternate procedure." The plumber was happy to learn that the Bureau still agreed with him.

Whereupon Washington exploded: "Don't use hydrochloric acid; it eats hell out of the pipes!"

Note: The second quotation on page 70 comes from Gertrude Stein's *Lucy Church Amiably.*

The answers to jargon

WILLIAM GILMAN

William Gilman was an editor for *Popular Science* and *Product Engineering*.

We are all guilty, more or less, of writing jargon. We are all guilty, too, of complaining about other people's jargon while we tolerate it in our own writing. Sometimes it sneaks past our guard. Other times we drag it in to brandish our knowledge—or hide what we don't know.

With jargon, as with other types of obscurity, we can blame much but not all of the trouble on "big words." Jargon can be ingratiatingly simple. *Metals Progress* magazine offers these examples of the kinds of statements that occur in research reports, and in parentheses explains what they really mean:

> While it has not been possible to provide definite answers to these questions . . . (The experiment didn't work out, but I figured I could at least get a publication out of it.)
>
> Three of the samples were chosen for detailed study . . . (The results on the others didn't make sense and were ignored.)
>
> It is suggested that . . . (I think.)
>
> It is clear that much additional work will be required before a complete understanding . . . (I don't understand it.)
>
> Correct within an order of magnitude . . . (Wrong.)

These are the tactics of fudging, a euphemistic word for dishonesty. But jargon varies—it can also be quite honest. Perhaps the saddest jargon of tongue or pen, and too common in the sciences, is that of the expositor who has meekly followed instructions.

For example, take a long breath and wade through this offering by a magazine:

> He explained that spatial vector electrocardiograph generalizes transfer impedance into a spatial continuum concept where it becomes represent-

From *The Language of Science: A Guide to Effective Writing* (New York: Harcourt Brace Jovanovich, 1961). Reprinted with permission of the estate of Eleanor E. Gilman.

able as a spatial vector point function very useful in determining the
sequence of the cardiac cycle temporally and spatially. An extension of
this theory, involving a multiple loop feedback dipole synthesizer, makes
it possible to compute automatically and instantaneously the optimal
vectorial point dipole representation of the heart as a current source.

This is hardly friendly, attractive prose. Even banks have learned that it
pays to come out of hiding—that big glass windows and open counters bring in
customers.

One thing is clear. The writer shied away from trying to translate. He
lifted jargon verbatim from a research report. And the scientist himself probably was not confused. He had substituted a rule for common sense—he was
dutifully using "precise" words. Quite likely these words were meaningful to
him. But they are also an undiluted outpouring of "big words" that would
repel even his fellow specialist. The pity of it is that many others—doctors,
instrument makers, computer engineers—would probably find this report
valuable if the scientist or writer had only let in some fresh air, some synonyms explaining that we have here a three-dimensional heart-tester that will
give instantaneous answers when used with a computer.

Another example has much simpler words, but they are dressed in the
garb of false humility:

In 1941 (Vol. 2, pg. 33) reviewer derived a formula for "Y." In reviewing
reviewer's article, author (Vol. 2, pg. 138) criticized reviewer's formula on
the grounds that it gave infinite "Z." In a later review of reviewer's article
. . .

And so on. This does not even have the doubtful virtue of supplying
bizarre words for the crossword puzzler. The writer here was obeying a typical
technical journal's warnings against the vulgarity of speaking in the first person. Result: a bewildering company of reviewers and authors that a simple "I"
and "my" would have sorted out.

One more example, this time at the other extreme—trying to popularize
but, in doing so, forsaking common sense. It's a publicity writer's announcement about some new industrial equipment. Jargon gives him a chance to play
cutely with words and you, the reader, abruptly find yourself riddling through
a panegyric to "low, low switchgear." What have we here? you ask. Might
there be some special meanings for *low*? Or is this a salesman screaming
bargains? When "low" and "lowest" become overworked by his competitors,
he outpromises them with his "low, low prices." If you care to quarry deeper
into the announcement, you discover that the first *low* refers to height of this
equipment. It won't bump its head against the ceiling. The second *low* refers to
low voltage. The equipment won't shock you.

Here, jargon was also standard ambiguity. Ordinarily, however, jargon is
not so much the double meaning of two-faced ambiguity as the frustrating no
meaning at all. It is the specialty word that, innocently or serving as intentional

fog, floors or vexes the nonspecialist; it is the technical or secret vocabulary that slams the door on everybody who isn't a fellow lodge member.

This concealment, where the mission is "revealment," is a costly kind of writing. Everybody who writes has something he wants to tell or sell, and he wants to be read by more, rather than fewer, readers. Remember that even if you are not selling hardware, you are selling your words—you are persuading the reader to read, and to continue reading. Even the fellow lodge member will flee as quickly as possible from the spatial-vector stuff quoted earlier.

Because it is so universal, like sin, jargon will always be with us, certainly in specialized writing. Much of the "dynamic" language of motivation research is jargon—the up-to-the-minute kind. And the "blameless" science, archaeology, speaks of its peaceful digging in the tongues of yesteryear—again, jargon.

How, then, can we curb this costly nuisance? How can we let most of the wind out of our billowing wordage? How, also, can we avoid being forced into a grim choice between jargon and illiteracy, between a usefully broad vocabulary and a primitively simple one? "Basic English" has been as ineffectual as Esperanto in offering any practical answer.

Any control of jargon must consider these realities:

a. There is no sure cure. Jargon is a universal fault. Merely haranguing against it is too much like the sermons of Calvin Coolidge's preacher, who was merely "agin sin."

b. More than any other writing fault, jargon must first be recognized before we can consider remedies. In this chapter we will see that it is often the mote in the other fellow's eye (and he sees it in your eye too).

c. It speaks in many tongues. Alexander Woollcott, fond of such nostalgic words as *wraprascal* and *tippet*, was as guilty of jargon as the much criticized technician or the semanticist with his "science of non-elementalistic evaluation" that considers the "neuro-linguistic and neuro-semantic environments" of an individual.

d. Don't expect the reader to know what he can't find in his standard dictionary, to recognize *Screwdriver*, *Countdown* and *Marstini* as vodka drinks of the Space Age. He can't be consulting all the lexicons of slang and jargon, and even they are neither complete nor up-to-date enough.

e. What is jargon to one reader may not be to another. Jargon can be uttered nonsense or precise label-words that science needs. These words have invaded chemistry, for example, in waves: the coal-tar derivatives that came in after Germany's monopoly was ended by World War I; the petrochemicals leading to new polymers and therefore new plastics after World War II. These name-words are needed. The sin of jargon is not so much in using "big words" as in flaunting show-off words, on the one hand, and deliberately esoteric words, on the other.

f. Universal sin is not easily legislated out of existence. Some countries try outlawing prostitution; others prefer licensing as a control. Writing, of course, is traditionally lawless. In a word-game like Scrabble you are restrained from using *syzygy* because there are only two *y*'s in the entire set of players' tiles, but nothing can prevent a typewriter from rattling off many *y*'s and *syzygies*. With jargon, then, the only hope is in moderation. The less jargon, the better; to practice this moderation, the best control is self-control.

g. When in doubt, cling to the reader. He's your salvation as well as your judge. Decide who he is—to what extent he shares your vocabulary—and then use only the technical and jargon words that he can be expected to understand or to decode easily. It's as simple as that.

The company you keep

You cannot swear off jargon effectively until you know just what it is. Otherwise you may think you are being asked to toss out culture and diction and learning. Also, we have seen that science does need its precise name-words. In our specialized age, the evolution from *insecticide* to *pesticide* and then *adulticide* makes some sense, too, even though *bug spray* would satisfy most readers.

An oddity about jargon is that it is so often condemned by practitioners of jargon. When writer A brandishes *existentialism*, writer B calls him *obfuscator*.

This blaming the other chap is common in science, and understandable when we realize that science is not so much a language as a phalanx of dialects. We now have writing by "computer engineers," "gyro engineers," "electronics generalists"; by "behavioral scientists," "human factors scientists" and even "human engineers." This only worsens what is bad enough already. A horticulturist and a physicist, both interested in radiation, complain that they cannot understand each other. And if the physicist happens to be a home gardener, he asks why the horticulturist gets lost in jargon instead of explaining why squash and pumpkin do not crossbreed. Moreover, if they are music-lovers, they share Debussy's disgust with the "fog of verbiage" under which music critics have buried Beethoven's Ninth Symphony.

Then, too, all the old-line sciences snicker at the tall talk that befogs writing in the "pseudo" and social sciences. Yet we know that both psychiatry and surgery, for example, have had to contend with impostors who learned the jargon and went ahead successfully to practice the profession.

Nevertheless, jeering at the other fellow has its value. When you learn to spot the other chap's jargon, you have taken the first step toward recognizing your own "parrot talk" and "baffle gab."

Jargon, of course, has many relatives. It stands midway between cant, which is first-degree murder of the language because it is deliberate, and the cliché, which is more often empty-headed meaninglessness. It was cant that made the author of *Tristram Shandy* choleric:

> Grant me patience, just Heaven!—Of all the cants which are canted in this canting world—though the cant of hypocrites may be the worst—the cant of criticism is the most tormenting!

Before science grew so big and voluble, philology and its subdivisions did most of the fretting about jargon and how to define it. On their tongues the word experts rolled the differences and similarities—jargon, lingo, argot, gibberish, and then the gobbledygook of bureaucracy and the weirdies of the rock-'n'-roll and beatnik clans. The experts generally saw close kinship between jargon and cant. This was unfair, because cant is often defined as the secret language of thieves and vagabonds. The jargonist is a thief only in the sense that he steals a reader's time. Nevertheless, it is manifest that jargon keeps rather bad company.

Sir Arthur Quiller-Couch derided "the Vanity Fair of jargon" and called it the doorstep to ambiguity. As another outspoken English scholar, Fowler, did later, he was flailing jargon in general rather than the technician's variety, and had fun linking it to prim prose—the elegance that substitutes "domicile" for "home," "perspiration" for "sweat," and "lady dog" for "bitch."

This, then, is the pretentious type of jargon. It betrays innocents into saying "between you and I" when "between you and me," though less ornate, happens to be correct.

But jargon is also the blah-blah of a politician who says he doesn't want higher taxes or a cabinet office when he actually does. Shortly after taking up his duties in Washington, a Secretary of Defense complained, "The language of legislation is spooky. I keep asking myself, 'What is it really saying?'"

Jargon is the bucolic affectation of the truck driver who says "tarpolean" when he knows it's "tarpaulin"; the member-of-the-club badge worn by cavalrymen, including educated West Pointers, who disdained the rest of the world by calling themselves the "calvary"; the speech of the baggy-pants burlesque comedian who gets a laugh from the yokels by pronouncing "connoisseur" as "conna-sewer."

But jargon is also the language of the science student who outgrows pig Latin and plausisounding nonsitalk—in favor of a spanking new amusement that will set him and his classmates apart from the rest. After making you guess, in vain, he explains that

$$\ln\left[\lim_{z \to \infty}\left(1 + \frac{1}{z}\right)^z + (\sin^2 x + \cos^2 x) = \sum_{n=0}^{\infty} \frac{\cosh y \sqrt{1 - \tanh^2 y}}{2^n}\right]$$

is the scientific way to say: One plus one equals two. Contrast that contrived equation with the simplicity of Einstein's literally earthshaking $e = mc^2$, which explains the power of a nuclear explosion.

Jargon can be the exhilarating slang of the barkers and midgets who work at carnivals. It can be the delightful nonsense (with pinpricks of meaning) of Lewis Carroll writing about Alice in Wonderland and of baseball's Casey Stengel testifying before a congressional committee.

But jargon can also be serious stuff—pomposity of the cultist and passwords of the secret-society member. The jingoist is usually a jargonist outfitted with a flag-waving set of catchwords; and Communist jargon has irritants of its own. Where is the difference when you weigh: deviationism, dogmatism, fractionalism, reformism, sectarianism, revisionism? All these, C. L. Sulzberger wrote in the New York *Times,* were Moscow synonyms condemning Marshal Tito's independent-mindedness.

Capitalism has its jargon, too, and critics such as Thorstein Veblen have had a merry time at the expense of soothsaying economists by mimicking their language with deadpan derision.

Education, of course, has its own jargon. An example is from a teacher's report on her youthful pilgrim's progress:

> He is adjusting well to his peer group and achieving to expectancy in skill subjects. But I'm afraid his growth in content subjects is blocked by his reluctance to get on with his developmental tasks.

(In other words, this lad is a good mixer, okay in the three R's but not much imagination.)

And the best jargon of all is the kind that can jeer at itself. Listen to this draftsman. He is explaining in doggerel that, to show how much he knows, "I have marked all my lines/ With mysterious signs/ That Einstein could never decode," and he goes on:

> Now my drawing is finished and printed,
> And I'm proud of its hazy design,
> For I know there'll be chaos and ulcers
> When it finally comes out on the line.
> And a feeling of pride starts a stirring inside,
> As my tracing is filed on the shelf,
> For my quest has been solved with a print so involved
> That I can't even read it myself.

Now let us dig a little deeper, more clinically, into this matter of detecting other people's jargon so that we can better recognize our own. Here's a glimpse at the language of two quite different technical fields.

Example 1—architecture. Paintings, clothes and buildings have this in common: They sell themselves to us through our eyes—they are designed. The fashion designers and painters have their jargons, of course, but neither of these fields is closely related to science. Their jargons are not deadly, but merely those of aesthetics.

Architecture, on the other hand, does incorporate much applied science, such as the physics of load-bearing structures and the chemistry of materials. And in architecture the floodgates of jargon open wide.

One illustration . . . [can be seen in] the three decades that R. Buckminster Fuller had to wait before the world understood and paid tribute to his "Dymaxion Principle" and the "tensegrity" (tension plus integrity) incorporated into his fantastically strong geodesic domes. He had evangelized for "energetic-synergetic" geometry and the distribution of forces in lattices made up of tetrahedrons. He had not explained that the dome is an exquisite balancing of the forces that try to make it collapse and the forces that try to tear it asunder; that this is the trick with which you could build a sort of thick-skinned tent in which the tautness is built right into the latticework of the skin—so that there is no need for a centerpole.

Another example shows how easy it is to note only the mote in the other fellow's eye. In a discussion of architectural jargon, a professor of architecture writes:

> The aspects of architecture embody so many factors that its definition has become almost unintelligible. It encompasses sociology, biology, physiology, aesthetics, engineering, space, decoration, etc., to the extent that architects attempting to write about it invariably engage in such mixed jargons that it results in an incomprehensible babel.

He then goes on to explain himself in such terms as "emanation" of a drawing, "disbalance," "cumulative field of reference," "dynamic equilibrium" and "prosaic agglomeration."

Certainly even rhapsodizing by John Ruskin tells us much more:

> For indeed, the greatest glory of a building is not in its stone, not in its gold. Its glory is in its age, and in that deep sense of voicefulness, of stern watching, of mysterious sympathy . . . which we feel in walls that have long been washed by the passing waves of humanity.

Example 2—medicine. This field, of course, is rife with jargon, partly because its roots are in the witch-doctor mumbo-jumbo that considers *toxicosis* always better than *poisoning*. Partly, too, because of the morale-building approach that gives the patient some pink pills for imagined ills and holds back the crisp truth about serious ones. As long as the prescription is written correctly, the patient is probably as proud of long-word ills as of plebeian ones; the misunderstandings are no more fatal than that between the young husband and wife who slept separately one summer because polio experts had been warning that the disease is spread by "intimate contact between people." The young couple understood "intimate" only in the prim-prose sense.

It is likely that medicine's jargon is most harmful to the doctors themselves. A specialist said as he came away from a research symposium: "I had to listen so hard to the words that I didn't have a chance to grasp the idea." And colleagues admitted nodding off during the blessed dark moments when slides were shown.

Yet medical matters can be described clearly—as a medical researcher like Hans Zinsser, and other writers such as Sinclair Lewis and Arnold Zweig have so brilliantly demonstrated.

The following passage is from Aldous Huxley's *After Many a Summer Dies the Swan*. This is a novel, but Huxley is a walking encyclopedia of science as well as an extremely competent writer, and his science fiction is as disciplined as any research report. In this passage, disregard his "style"; instead, note the expositional clarity—how words that might be jargon are quickly unfolded for you. At the beginning you know little about carp and probably less about the sterols; at the end you are ready to live as long as Methuselah:

> Those sterols! (Dr. Obispo frowned and shook his head over them.) Always linked up with senility. The most obvious case, of course, was cholesterol. A senile animal might be defined as one with an accumulation of cholesterol in the walls of the arteries. . . . But then cholesterol was only one of the sterols. They were a closely related group, those fatty alcohols. . . . In other words, cancer might be regarded, in a final analysis, as a symptom of sterol-poisoning. He himself would go even further and say that such sterol-poisoning was responsible for the entire degenerative process of senescence in man and the other animals. What nobody had done hitherto was to look into the part played by fatty alcohols in the life of such animals as carp. That was the work he had been doing for the last year. His researches had convinced him of three things: first, that the fatty alcohols in carp did not accumulate in excessive quantity; second, that they did not undergo transformation into the more poisonous sterols; and third, that both of these immunities were due to the peculiar nature of the carp's intestinal flora. What a flora! . . . In one way or another, in combination or in isolation, these organisms contrived to keep the fish's sterols from turning into poisons. That was why a carp could live a couple of hundred years and show no signs of senility.

Less jargon—your guide

Now let's focus on your own writing. You think you can recognize jargon. You want to cut it out of your sentences and paragraphs, or at least cut it down. But how?

1. *Preventives come first.* Beware of infection by the other person's jargon and you will save time on cures. Here are some ways to avoid infection.

a. Don't parrot the instructor's bigger words. They may get you a college degree, but professional life requires writing that is more than regurgitation. Jargon, for example, doesn't impress a capable boss. He's too busy, and wants to pick your brains in a hurry. He doesn't equate complex writing with complex thoughts and, in fact, is suspicious of thoughts that cannot be simplified to the length of, say, Lincoln's Gettysburg Address.

b. Spurn the words that bother you in your reading—they may bother your own readers. For example, just as too many isms, each requiring its own political party, can paralyze the functioning of a government, so can the spinning off of too many -ologies throw our language into bedlam. Let metaphysics insist on its ontology, but need our writing stagger under all these: behavioristic psychology, Gestalt psychology, normic psychology, structural psychology, functional psychology, act psychology, dynamic psychology, reflexological psychology?

c. If you are editing or rewriting the work of others, translate the jargon when you can; when you can't, call for something simpler from the author. Otherwise, you are an accessory after the fact. As editors know, this can lead to troublesome bickering. The best cure is to let the author take his jargon elsewhere. Of course, before you challenge fact or jargon, be sure of your ground. A garbage grinder, for example, may prefer the sweet smell of another name. A magazine editor learned this from a plaintive letter sent in by a manufacturer:

We prefer that you call it correctly as "Industrial Disposal Unit." Garbage grinder connotes shoddy appearance, etc. Certainly, any machine capable of disposing of "Top Secret" documents should not be called a garbage grinder.

2. *Who's your reader?* Delay writing until you have decided who the reader, or class of readers, will be. If you don't know, you are still too hazy about the subject itself.

Is it to be a proposal to the boss? That isn't identification enough. Is he strictly a management man, does he have a smattering of science, is he a fellow technician? Or is it to be a pass-along proposal that will be read by all three types of person?

If your paper is for fellow specialists, the jargon content can go up. But again, remember that people in promotion and advertising may have to read it, too, and they cannot translate what they cannot understand.

If you are writing for a magazine, who are its readers? If for students, are they at graduate or freshman level? And what of the so-called general public? It turns thumbs down quickly even on a literate writer if he strays too far from words like *home* and *income*.

When picturing the audience, remember: (a) Clarity brings its own reward—the less jargon, the wider the potential readership; (b) if you really prefer a narrower audience, don't be misled by specialization. Writing for "farmers" isn't narrow enough—the applegrower has one lingo, the poultry-man another. Similarly, your comrade technician isn't your twin in training and interests.

3. *Simplification isn't that simple.* Don't let the preaching against jargon panic you into such lisping simplicity as "A river pump house pumps water to the water treatment plant" and "He was a self-educated man educated by himself."

And by all means don't be panicked into popularization that falls on its face. When the United States Weather Bureau introduced its Discomfort Index, that term was simplicity itself compared with the attempt by a newspaper to explain it:

> . . . The relationship between such variables is found by computing a third measure. This is a magnitude so related to the magnitudes of temperature and humidity that to values of these measures there correspond values of the third measure. This third measure is called a function, a measure of relationship. . . .

4. *If you don't know, say so.* Is your chain of facts secure? If links are missing, you have this choice: (a) Ignore the gaps—but this leaves you open to being called slipshod, or worse; (b) delay the final draft until you can supply the missing facts, or bypass the trouble area and limit your writing to what you do know about; (c) come right out and say "I don't know." Truly great scientists frequently exhibit such humility. It can also be very refreshing. Thus, in a nonpompous article in *Scientific American*, Robert T. Wilson, the physicist, discusses Soviet particle-accelerators:

> Veksler . . . envisages a small bunch of ions in a plasma. . . . These waves are to act together coherently to give an enormous push to the ions being accelerated. If this is not clear to the reader, it is because it is not clear to me. The details have managed to escape most of us because of a linguistic ferrous curtain, but Veksler speaks of the theoretical possibility of attaining energies up to 1,000 bev. . . .

5. *Ordinary words are respectable.* Don't let very simple words cause you shame. Note that Wilson, above, used *bunch*. Other topnotch technicians are similarly unafraid of writing like human beings. If nuclear writing can be handled this way, why not that of the applied sciences? For example, medicine—which concerns everybody—might well emulate agricultural science, which also deals with quite ordinary people. Agronomy, horticulture, animal genetics—these have ponderous jargons of their own. But listen to the homely language of a typical county agent talking to his farm flock over the radio, or read the yearbooks issued by the United States Department of Agriculture. Some of the most capable writing by experts and editing by experts can be found in these technical annuals.

The researchers in agriculture, fighting the gloomy predictions of Malthus, learned long ago to clarify their writing or let others do it for them; that a discovery is incomplete until fellow technicians, and eventually the farmers, can apply it.

6. *Common cure—a synonym.* Whenever you suspect jargon is creeping in, call on substitute words that are easier. . . . In general, simplicity of the word should depend on simplicity of the reader. New York's radio station WNYC, addressing all kinds of people, was smart enough to use "the coming bookkeeping year" instead of "fiscal year"; this avoided confusion in listeners' minds between fiscal and calendar years. Experts, too, are grateful for simpler words, for each time the chemist says "nylon" instead of giving the full molecular name of the polyamide.

7. *But some jargon is necessary?* If so, rush in to explain it to the reader. This can be done in several ways. A shortcut method is the parenthetical one. Be quite formal, use jargon if you must, but immediately define the word. The New York *Times* does this frequently and nicely, with the synonymous material flanked by parentheses or dashes. Thus:

The volume, amounting to 417 lots (50,000 pounds to a lot) was the second highest this year.

The metal is as light as magnesium alloys, yet its modulus of elasticity— or stiffness—is about three and one-half times that of steel.

The same method works handily when writing for a more technical audience. Thus, from the Beckman Instrument Company, an announcement about its D2 Oxygen Analyzer says:

. . . Oxygen is unique among gases in being strongly paramagnetic (attracted into a magnetic field). Other gases are, with few exceptions, slightly diamagnetic (repelled out of a magnetic field). . . .

Obviously, to be effective, the explanation must explain. In the following, from *Scientific American*, the reader either understands or doesn't understand precisely what is meant—the explanation (which I italicize) merely adds bothersome words:

The rate of change of the orbital period of a satellite, *that is, the rate of change in the time it takes to fly its orbit,* is directly proportional to the density of the atmosphere through which it passes.

8. *Comparisons are practical.* Another effective way to make jargon more palatable is to put similarity to work. An example is the "For example . . ." method so useful in all expositional writing.

For example, the atom was formerly likened to an infinitesimally tiny billiard ball—a hard little atom. But later, when the concept was revised to include spinning electrons, the student was asked to think of a tiny planetary system.

Or take mathematics—almost as much of a terror to most technicians as it is to laymen. A mathematician with a new concept about finite groups was asked by a newspaper reporter to explain it. His first answer was that this could not be done simply enough for the layman. But he then did it very effectively in terms of wallpaper patterns.

This method, however, requires considerable caution and common sense. The horrors of a mixed metaphor must be avoided. The analogy must be sharply to the point—or the reader's mind is sent wandering off the subject. And it's hardly helpful to have the mitosis of a human's cells likened to the mitosis of a guinea pig's cells—if the reader doesn't know what mitosis means.

Allow the example to be lengthy only if it is itself an integral part of the exposition. And don't reduce the reader's mental stature to that of a child. A New York *Times* story, in the course of reporting a speech by a university president, quoted this sentence, "Today, our specialists, as they study man and his behavior, are all too frequently like the blind men studying the elephant." At this point the reporter, evidently disbelieving that *Times* readers are fairly literate, launched into a parenthetical paragraph of 85 words that retold the familiar blind-men-and-elephant story.

9. *This is the smoothest way.* The good craftsman, like Aldous Huxley, quoted earlier, can have his cake and eat it too. He throws in troublesome words when necessary but quickly explains them with what might be called built-in paraphrasing, which advances the exposition while explaining the jargon. Difficult word and explanation come so closely together that the reader feels no pain, hardly realizes what has been done for him.

Let us work through a step-by-step example. The problem I am about to describe actually exists. Someday it may be solved. I am imagining that day has arrived and I want to announce the triumph. If I were writing a typical research paper, I would begin something like this:

The literature contains many references to a problem that has been important in silviculture for the last fifty years. . . .

And so on, until I come in pages later with what it is I want to announce.

An unappetizing beginning? Yes. But clarity is our main concern here. Focus on the word *silviculture*. How many readers can I assume would have a roughish idea of its meaning? Worse, how many would know its precise meaning? Very well then, I can conquer jargon this way:

The literature contains references to a problem that has been important in silviculture for the last fifty years. Silviculture is defined as the art of producing and caring for a forest. . . .

With this interruption for definition I have obeyed the injunction to explain any jargon. But at considerable expense. The unappetizing beginning has
become even more deadly, and I have wasted space just to give a definition.

Yet this *silviculture* is a word I really want to use throughout the article; I
also believe sincerely that if the reader doesn't know the word, it's high time
he learn it.

Now let's try the technique by paraphrasing. If I insist on beginning with
a review of the literature, this will at least purge the jargon (the difficult word
and the explaining words are in italics):

The literature contains many references to a *forest-growing* problem that
has been important in *silviculture* for the last fifty years. . . .

Or, the same, but with dullness thrown away:

Science is now certain that the American chestnut *tree* can be brought
back into our *forests*. Government *silviculturists* are successfully *growing thousands* of chestnut *seedlings*. These are identical in species with
the towering trees that perished when a mysterious blight began attacking them fifty years ago. . . .

Or, the same, extremely popularized:

The deadliest killers of America's *trees* has now been licked. Fifty years
ago, *silviculturists* watched helplessly as a mysterious blight swept
across the country, turning *forests* of lordly chestnut trees into companies
of gray ghosts. Five years ago, science found in coal tar a chemical that
was just as deadly against the blight. A month ago, *silviculturists* of
another generation jubilantly began *planting new forests* of chestnuts. . . .

Or, very simply, merely telling somebody about something I do on my
farm:

I like *silviculture*. *Growing forest trees* is more than a hobby with me.

Here, I have had different audiences in mind but the method is always the
same. I define the word as I move along.

Another example. I cannot assume that all readers of this book are
steeped in the jargons of grammar and semantics. But I do assume that most of
you have had a college education or will have. Therefore I feel no need to
define such words as *antecedent* and *adjective*. But in the chapter on new words
I was much more doubtful about *neologism*. Yet I felt you should know this
word. So I hastily defined it, not by pedantically saying, "A neologism is . . ."
but by letting the act of definition do some other work for me. I had been

discussing new words. Then I went on, "Even if the neologism has been broken in a little, enough to be found in some dictionary . . ."

All of the preceding how-to's can be lumped together as follows: Don't expect the reader to supply the explanation. He's too busy or doesn't know how. If you cannot explain, change the subject, or aim for another kind of reader.

While I was at lunch one day with some visiting engineers from the USSR, the conversation turned to translations. I asked if these foreigners had read Vance Packard's *The Status Seekers*. They said they expected to—the book was being translated into Russian. And what about *The Hucksters*? I asked. They nodded with comprehension—yes, they had heard about it. But, they added, it wouldn't be translated into Russian. Why? Their answer was convincing. The USSR had status-seekers too—so the Packard book would make sense. But the concept of free-enterprise advertising, of huckstering, defied translation into terms that Russians could understand. So Frederic Wakeman's book on hucksters would be jargon—to Russian readers.

A rationale for the use of common business-letter expressions

J. HAROLD JANIS

J. Harold Janis is Professor Emeritus of Business Communication at New York University; a third edition of his textbook *Writing and Communication in Business* was published by Macmillan in 1978.

*T*he aim of this paper is to provide a rational basis for the use of stock words and phrases in business letters. This aim follows the hypothesis that such expressions can, under controlled conditions, serve the needs of business and at the same time avoid criticism on stylistic grounds. Resolving the question of usage is important in order to stop the great waste of time in promoting language values that are often at odds with the business culture. It is also important if the constantly increasing volume of routine correspondence is to be handled in an effective and acceptable way by computer or other systematic means.

The number of pejorative terms used to describe common business-letter expressions suggests the difficulty of treating (and viewing) the subject dispassionately. Adjectives like *stereotyped, stilted, hackneyed,* and *trite,* and nouns like *clichés, jargon, bromides, businessese,* and *gobbledygook* have so colored our attitudes toward certain types of words and phrases that any attempt to treat the offending usages on a reasonable basis risks skepticism and antagonism. To eliminate, or at least reduce, the affective connotation, we shall hereafter use the abbreviation CBE to signify a common business-letter expression.

From *The Journal of Business Communication,* 4 (October, 1966). Reprinted by permission of the American Business Communication Association, publishers of *The Journal of Business Communication,* and the author.

Factors favoring CBE's

This author has previously shown that the use of certain common business expressions is grounded in the demands of the business situation.[1] The organizational factors favoring the use of CBE's may be summarized as follows:

1. Letter writing in business is highly repetitive. The same kinds of situations occur again and again, providing the writers with ample precedent for both the substance and phrasing of the letters. If this were not so—if each situation were unique or had to be treated as if it were—business correspondence on a large scale would take a disproportionate amount of the organization's energies.

2. CBE's increase efficiency by providing ready clues to rhetorical patterns and consequently reducing the uncertainties of expression. For example, given the initial phrase *with reference to your letter of,* the writer is easily able to begin not just one letter, but a great many. When the writer is required to begin the letter in less redundant (less predictable) fashion, e.g., *We don't know how we mislaid your order,* he is left completely to his own devices and, furthermore, cannot rely on such phrasing to provide any assistance in composing other letters.

3. CBE's help to reduce the uncertainties of response. From the point of view of the writer's superior, any new language treatment increases the risk of message failure, including misunderstanding and legal liability. He is therefore inclined to reject originality in favor of phrasing proved by experience. Conformity is thus enforced by the threat of non-acceptance.

4. CBE's permit the correspondent to be impersonal in the many instances when he does not have any personal involvement in his subject or considers it desirable to avoid personal responsibility. In many instances the correspondent has had only a small part in the transaction about which he is writing, or he may be writing for some other person's signature. The tradition of the particular company or department for which the writer works may also encourage his self-effacement. The use of CBE's like *receipt is acknowledged, the undersigned,* and *our records indicate* becomes more understandable in the light of this explanation.

Reconciling conformance with effectiveness

If, as it is indicated, the business environment nurtures conformance in language rather than freshness and originality, how is the use of CBE's reconciled with the need for effectiveness? Several explanations may be offered. First,

---

[1] J. Harold Janis, "The Writing Behavior of Businessmen," *Journal of Communication,* 15 (June, 1965), pp. 81–88.

despite the emphasis on the human relations function—and the need for individuality that goes with it—the task function is paramount in a huge volume of business correspondence, certainly in the routine situations most likely to be handled by subordinates. And whatever the prevalence of CBE's, the routine business-letter tasks apparently do get done.

Second, the factors that make CBE's desirable for the writer also make them desirable for the reader. When similar situations keep arising between the two parties, the reader's response to the language is conditioned by his experience with it. Changing the language then becomes analogous to changing the color or design on a can of beans. Whatever the improvement, there is at least temporary disorientation.

Third, the value the reader places on the distinctiveness of a business letter can easily be overestimated. Apart from the substance, which is of the most immediate concern to the reader, a number of other factors contribute to the total impression the letter makes. These include its physical make-up, its timeliness, the rapport with the signer, and the reputation of the source.

Finally, the balance between an efficient style and effectiveness is preserved by what we shall call the fraction of composition. Following Schramm's concept of the fraction of selection (the attention paid to a communication)[2] the fraction of composition may be represented as follows:

$$\text{Fraction of Composition} = \frac{\text{Expected Reward}}{\text{Expected Energy Required}}$$

As the formula signifies, the more meaningful the anticipated response to the letter, the more available energy the writer or organization expends. Thus routine letter situations which have been adequately (if not triumphantly) handled in the past by CBE's, provide little incentive for change. On the other hand, unique situations that promise greater rewards (or greater penalties) provide greater incentive for individual treatment. In the simplest terms, the language of a letter gets all the emphasis it warrants, consistent with the proficiency of the writer and the competition for his time and energy.

Stylistic considerations

We have yet to deal with the stylistic or esthetic objection to CBE's. This objection appears to be based on the loss of individuality that occurs whenever clichés are used. Thus *cliché* becomes the operative word. According to Beckson and Ganz, a cliché is "a timeworn expression which has lost its vitality and

---

[2] Wilbur Schramm, "How Communication Works," in *The Process and Effects of Mass Communication*, ed. by Wilbur Schramm (Urbana, Ill.: University of Illinois Press, 1960), pp. 19–20.

to some extent its original meaning."[3] Margaret Nicholson defines clichés as hackneyed phrases that "have acquired an unfortunate popularity and come into general use even when they are not more but less suitable to the context than plain speech."[4] Fowler says that "cliché means a stereotype; in its literary sense it is a word or phrase whose felicity in a particular context when it was first employed has won it such popularity that it is apt to be used unsuitably and indiscriminately."[5]

The test of a cliché, then, is not that it is common, but that it is both common and unsuitable for its context. On this point Fowler offers some elaboration:

> The word is always used in a pejorative sense, and this obscures the truth that words and phrases falling within the definition are not all of a kind. There are some that deserve the stigma. . . . There are others that may or may not deserve to be classed with them; that depends on whether they are chosen mechanically . . . or are chosen deliberately as the fittest way of saying what needs to be said.[6]

Adding that "the enthusiasm of the cliché hunter is apt to run away with him," Fowler quotes from J. A. Spencer that "the hardest working cliché is better than the phrase that fails" and "journalese is best avoided by the frank acceptance of even a hard-worn phrase when it expresses what you want to say."[7]

What has been said of clichés can be said of CBE's. The words and phrases falling within the definition are not all of a kind. What matters is their appropriateness in the context in which they are used. To give an example from commercial writing, *we wish to advise you* would be stereotyped in a letter acknowledging an order, but not necessarily inappropriate in a legal notification. As the example suggests, an important clue to the aptness of CBE's can be found in the stratification of style, which is characteristic of organizational correspondence. In this respect, four levels of usage may be noted (Fig. 1):

Official. This style is used in correspondence of a legal or quasi-legal kind, including claims, requests, notifications and acknowledgments. In addition to its immediate utility, the correspondence also serves as a formal record of a transaction. The style is highly impersonal and characterized by stilted expressions that would be considered out of place in other kinds of writing. Such expressions arise out of either the desire to preserve the legal tone (and force) of the message or the need for exact reference. They include *pursuant to, advise* (in the sense of "inform"), *above-captioned, therein, thereto, thereof,*

[3] Karl Beckson and Arthur Ganz, *A Reader's Guide to Literary Terms* (New York: Noonday Press, 1960), p. 29.
[4] Margaret Nicholson, *A Dictionary of American-English Usage* (New York: Oxford University Press, 1957), p. 88.
[5] H. W. Fowler, *Modern English Usage* (New York: Oxford University Press, 1927), p. 90.
[6] *Ibid.*, pp. 90–91.
[7] *Ibid.*, p. 91.

Fig. 1. Levels of Business-Letter Usage

OFFICIAL

As requested in your letter to us of November 28, 19___, all of the $200,000 principal amount of your Corporation's 5% Registered Debenture Bonds due 1970 called for redemption on December 1, 1965 at the principal amount thereof and three months' accrued interest have been presented to us and paid. After such redemption, there remains outstanding $850,000 principal amount.

With further regard to your claim covered by our File No. 1467B, we regret to advise that after due consideration of the circumstances surrounding the accident, the Claim Board has reached the conclusion that we are not justified in making payment in this instance.

FORMAL

This will serve to introduce Miss_____, who has been in our employ since July 2, 1964 as a typist in our Credit Department. Miss_____, an excellent worker, is leaving our Company on March 22, 1966 to reside in Puerto Rico. Mr. William_____ of the New York Hilton Hotel, one of our valued customers, said you might know of some job opportunities for Miss_____.

With regard to your recent inquiry, Mr. John_____ has been numbered among our depositors since this office opened ten years ago. Mr._____ maintains a satisfactory personal account on which balances average consistently in four figures, and he is also known to us in connection with several substantial business accounts. On the basis of our experience we have no hesitancy in recommending Mr._____ as deserving of the usual courtesies.

INFORMAL

We have your letter of September 25 requesting information concerning 300 shares of Blanko Corporation that you sent us for transfer.

In order to help us identify the item, will you please note on the carbon copy of this letter the name of the person to whom the stock was to be transferred and also the certificate numbers of the shares you sent us. We would appreciate your writing on the enclosed carbon of this letter and returning it to us in the envelope we have provided.

I am sorry about the lost check. If you will complete the enclosed Affidavit, I will be happy to send you our official check for $256.70.

COLLOQUIAL

My face is red over the delay in sending you and Ed the tax information I promised. Life has been hectic but not that hectic.

Lunch Friday will be just fine. See you at the Lawyers Club at 12:30.

therefor, due to (as a preposition), *subsequent to, held in abeyance, duly,* and *pending receipt of.*

Formal. The formal style serves many purposes calling for an impersonal but not legalistic treatment. The communications may include letters of introduction and recommendation, credit references, early collection letters, and correspondence on routine matters with agents who are not emotionally involved in the situations. In addition to the use of the impersonal *we* (for *I*), *the writer,* and *the undersigned,* the style is also characterized by polite expressions like *kindly* (for *please*) and *at your earliest convenience,* and such formal but wordy expressions as *in the amount of, with reference to, in connection with, in the event that,* and *our records indicate.*

Informal. The informal style is suitable for most business letters. It is characterized by a personal interest in the reader and contains a considerable amount of affective language, including such expressions as *glad, pleased, sorry,* and *appreciate.* Thus the human relations function, while not necessarily absent in the official and formal styles, is more prominent—sometimes predominant—here. Informality is enhanced by the personal pronouns *I, we,* and *you,* active verbs, and natural (unstilted) phrasing.

Familiar. When writer and reader know each other quite well, and especially when they meet socially or address each other by their first names, the familiar style may be appropriate. It is marked by colloquial expressions, clipped sentences, and an occasional touch of humor. Although any form of stiltedness is incongruous in this style, such CBE's as *with all good wishes,* and *kindest regards* are quite common.

Guidelines for the use of CBE's

Given these classifications of usage, we are in a better position to determine the rightness of CBE's in particular situations. We are also better able to form judgments about the use of CBE's generally:

1. The chief objection to CBE's in business letters is that they are "not natural." Obviously, such a criticism cannot apply to the official and formal styles, which imply a degree of stiltedness. When the criticism is applied to letters for which the informal style would be best suited, the substitutes usually recommended are not original modes of expression but other CBE's that will pass as natural. The contrasting phrases at the top of the next page, for example, have been compiled from several well-known texts on letter writing.

2. Much of the prejudice against CBE's is really a prejudice against the use of official or formal language in situations that would be more appropriately handled by informal language. This is legitimate criticism.

STILTED CBE's	NATURAL CBE's
at your earliest convenience	as soon as you can[8]
attached herewith	attached is[8]
This is to acknowledge	thank you for[8]
awaiting your further orders	we shall expect to hear from you[9]
under separate cover	separately[9]
as per instructions	following your instructions[10]
pursuant to your request	as you requested[10]
we thank you kindly for your letter	thank you for your letter[10]

The following letter, for example, is needlessly stilted, as the writer himself seems to sense in the postscript:

Gentlemen:

We are carrying you on our mailing list as follows: . . . Will you please return this notice with any corrections shown thereon which should be made. If the above information is correct, we would appreciate advice from you to that effect so that our records will be complete.

Your truly,

P. S. In other words, we would appreciate having your correct mailing address.

3. Any excesses in the use of CBE's are bad style. Thus while *thank you for* is suitable in informal usage, it palls when it is used without discrimination to acknowledge incoming letters. Similarly, the stiltedness characteristic of formal or official usage becomes a travesty in a passage like the following:

In response to your letter of March 26 which was in reply to our letter of March 24, which was in reply to yours of March 16, we are enclosing herewith the photostatic copy of our check No. 60432 dated February 26, 19—, payable to (Name) in the amount of $6,455.54.

4. Some CBE's are hard to justify in any circumstances. This judgment would apply to trade jargon which is not adapted to the reader and for which there are reasonably adequate substitutes. It would also apply to archaic or stilted or pretentious expressions which have no redeeming advantage. Illustratively, such expressions might include *and oblige, we remain, as per, your goodselves,* the *instant matter, favor* (for letter), and *summarization* (for *summary*).

5. Regardless of the level of usage, expressions that are stylistically undesirable in some instances are defensible and even appropriate in oth-

[8] Charles C. Parkhurst, *Business Communication for Better Human Relations*, 7th ed. (Englewood Cliffs, N. J.: Prentice-Hall, 1966), pp. 19–20.
[9] Mona Sheppard, *Plain Letters* (New York: Simon & Schuster, 1960), p. 99.
[10] Walter K. Smart, Louis W. McKelvey, and Richard C. Gerfen, *Business Letters*, 4th ed. (New York: Harper & Row, 1957), pp. 36–37.

ers. For example, a certain credit department recommends to its correspondents a guide letter in which the following form appears: ". . . (company) is engaged in the (manufacture) of. . . ." One might contend that "the company manufactures" is simpler and therefore better. From the credit department's point of view, however, the phrase *engaged in* is a clear reminder of the kind of information that is to be given at that point. The word *manufactures* following the word *company* would not suggest as well to the correspondent the possibility of such substitutions as *imports*, *retails*, or *jobs*. It is also evident that the noun construction, which gives the common name of the company's activity, is less awkward than the verb construction in some instances (as "the company jobs") and is better suited to the purpose of the letter.

In summary, (1) use of the common business expressions (CBE's) most often associated with routine business letters is a normal consequence of the business culture, (2) such use is not inconsistent with the need for effectiveness, (3) CBE's are not necessarily clichés and may therefore be exempt from criticism on esthetic grounds, and (4) the criterion of good usage in a CBE is its appropriateness to the business situation and to the formality of the context.

The plain english revolution

ALAN SIEGEL

Alan Siegel, a leading consultant in the Plain English movement, is the President of Siegel & Gale, a firm specializing in language simplification.

LOUIS XVI: C'est une grande révolte?

DUC DE LA ROCHEFOUCAULD— LIANCOURT: Non, Sire, c'est une grande révolution. (When the news arrived at Versailles of the Fall of the Bastille, 1789)

*T*hree-and-a-half years ago I wrote an article for this magazine on the movement away from legalese toward forms and documents that can be understood by ordinary citizens.* At that time, the movement consisted of isolated revolts against legalistic gobbledygook and bureaucratese. A handful of business documents—mainly insurance policies and loan notes—had been simplified by forward-looking companies. In government, the scattered advocates of clear communication shared the lament of Alfred Kahn, then head of the Civil Aeronautics Board: "There's nothing I can do but cry. I feel so lonely and futile."

Now the plain language movement is becoming a revolution. (See the box on page 102 for a list of important advances.) But the Bastille has not fallen, yet. Part of the legal community remains resistant to change. Some supposed practitioners of Plain English have confused simplicity with simplemindedness. And the greatest obstacle has yet to be fully overcome—a misunderstanding of what Plain English is all about.

That misunderstanding was dramatized when New Jersey's Plain English Bill, passed by the legislature, expired in February 1980, through a pocket veto by Gov. Brendan Byrne, who said: ". . . although regulating the protection of consumers is a worthy goal, legislating the style of a society's prose is another thing."†

* "To life the curse of legalese—Simplify, Simplify," *Across the Board*, June 1977.

†Rewritten and reintroduced in June, the Bill was passed again. This time Gov. Byrne signed it (on October 10th), out of deference to its "unusually large" number of sponsors—65 of the Assembly's 80 members.

Reprinted from *Across the Board*, 18 (February, 1981) by permission of the Conference Board, the editors of *Across the Board*, and Alan Siegel.

But the movement has nothing to do with "legislating the style of a society's prose." We are not trying to turn English into one-syllable words, or to translate Saul Bellow into baby-talk. We are not even trying to do away with professional jargon, though that is a tempting target. Let lawyers talk to lawyers, or accountants to accountants, as they please. So long as they understand one another, that is just fine.

Well, what is the Plain English movement all about? *Its aim is to make functional documents function, whether they are put out by business or by government. If a consumer is expected to abide by a formal document—an insurance policy, a mortgage, a lease, a warranty, a tax form—then the consumer should be able to understand the document.* Carl Felsenfeld, vice president and counsel for Citicorp N.A., expresses the concept in legal terms: "There is growing dissatisfaction with contracts where consumers merely 'sign here' and can't, under any reputable system of contract law, be deemed to have agreed to all the printed verbiage." In social terms, I would say that people are learning a new right: the right to understand.

It's been argued that the fault lies with consumers, or with the educational system that does not train them to cope with legalese, or accounting, or insurance terminology. But that argument is neither reasonable nor realistic. In fact, documents meant for consumers have been made much more difficult than they really need to be. Redressing the balance—making sure that the documents consumers are expected to understand are made understandable—is a matter of simple fairness, and simple efficiency. In the long run it's in the interests of business as well as the consumer.

Five years of experience have proved that consumer contracts and forms of all kinds can be made much more understandable without sacrificing legal effectiveness. The very first plain language loan note, which I helped write for Citibank of New York, was introduced in 1975. Since then, Citicorp counsel Felsenfeld notes, "We've lost no money and there has been no litigation as a result of simplification."

That loan note remains a good example of just what plain language means. Among other things:

- A personal tone was used throughout—"I" and "me" rather than "the undersigned" or "Borrower," "you" and "yours" rather than "the Bank."

- Language was radically simplified. For example, "To repay my loan, I promise to pay you . . ." instead of "For value received, the undersigned (jointly and severally) hereby promise(s) to pay . . ."

- When unfamiliar terms couldn't be eliminated, we added explanatory phrases. For instance, ". . . if this loan is refinanced—that is, replaced by a new note—you [the bank] will refund the unearned finance charge, figured by the rule of 78—a commonly used formula for figuring rebates on installment loans."

- We shortened the sentences wherever possible and even used contractions ("I'll pay this sum . . ."). To enhance clarity, we chose active instead of passive verb forms where possible.

- Improvements in design included the use of larger (12-point) type printed in green on light brown stock. Compared with the previously intimidating format, this visually appealing approach suggests immediately that the document is supposed to be read.

The Citibank note taught us a fundamental lesson about simplifying language. Consumer contracts have traditionally been adapted from mercantile contracts, which feature verbose protective clauses accumulated over the years in an effort to cover all possible contingencies. Many of these provisions have no practical value in the consumer marketplace. So the first task in simplifying a document is not rewriting it in Plain English, but identifying clauses taken from commercial contracts that can be eliminated from the consumer contract without jeopardizing its validity. The secret to doing this lies in analyzing actual business experience to see which provisions are really being used.

In the case of the Citibank note, the provisions describing the lender's protections in case of default proved to be the biggest challenge. The traditional note listed a string of contingencies more than 180 words long:

Before

In the event of default in the payment of this or any other Obligation or the performance or observance of any term or covenant contained herein or in any note or other contract or agreement evidencing or relating to any Obligation or any Collateral on the Borrower's part to be performed or observed; or the undersigned Borrower shall die; or any of the undersigned become insolvent or make an assignment for the benefit of creditors; or a petition shall be filed by or against any of the undersigned under any provision of the Bankruptcy Act; or any money, securities or property of the undersigned now or hereafter on deposit with or in the possession or under the control of the Bank shall be attached or become subject to distraint proceedings or any order of process of any court; or the Bank shall deem itself to be insecure, then and in any such event, the Bank shall have the right (at its option), without demand or notice of any kind, to declare all or any part of the Obligations to be immediately due and payable. . . .

But analysis of Citibank's business experience disclosed that the typical consumer loan transaction needs only one event of default—failure to pay. With one additional protection added, the following replaced all the fine print above:

After

Default I'll be in default—
    1. If I don't pay an installment on time, or
    2. If any other creditor tries by legal process to take any money of mine in your possession.

A number of insurance companies, whose product, after all, is words on paper, continue to get good marks for simplification. The Massachusetts Savings Bank Life Insurance (SBLI) Whole Life Policy used to begin like this:

> IN CONSIDERATION OF THE APPLICATION for this policy (copy attached hereto) which is the basis of and a part of this contract and of the payment of an annual premium as hereinafter specified for the basic policy as of the Date of Issue as specified herein and on the anniversary of such date in each year during the continuance of this contract until premiums have been paid for the number of years indicated in the POLICY SPECIFICATIONS or until the prior death of the Insured . . .

By contrast, the cover of their simplified policy begins with this message:

> Please take the time to read your SBLI policy carefully. Your SBLI representative will be glad to answer any questions. You may return this policy within ten days after receiving it. Deliver it to any SBLI agency. We'll promptly refund all premiums paid for it.

The insurer's promise to pay the consumer used to be phrased in these forbidding terms:

> The Bank Hereby Agrees upon Surrender of this Policy to Pay the Face Amount Specified Above, less any indebtedness on or secured hereunder . . . upon receipt of due proof of the Insured's death to the beneficiary named in the Application herefor or to such other beneficiary as may be entitled thereto under the provisions hereof, or if no such beneficiary survives the Insured, then to the Owner or to the estate of the Owner.

In Plain English, policyholders can understand what they are buying:

> We will pay the face amount when we receive proof of the Insured's death. We will pay the named Beneficiary. If no Beneficiary survives the Insured, we'll pay the Owner of this policy, or the Owner's estate.
>
> Any amount owed to us under this policy will be deducted. We'll refund any premiums paid beyond the month of death.

Perhaps the most ambitious simplification program so far has been that undertaken by the St. Paul Fire and Marine Insurance Company, of St. Paul, Minnesota. In 1975 they became one of the first large insurers to begin simplifying their policies. As a pilot project, the company produced an easy-to-read "personal liability catastrophe policy" that used personalized examples in colloquial English:

Before

a. Automobile and Watercraft Liability:

    1. any Relative with respect to (i) an Automobile owned by the Named Insured or a Relative, or (ii) a Non-owned Automobile, provided his actual operation or (if he is not operating) the other actual use thereof

is with the permission of the owner and is within the scope of such permission, or

2. any person while using an Automobile or Watercraft, owned by, loaned or hired for use in behalf of the Named Insured and any person or organization legally responsible for the use thereof is within the scope of such permission.

After

We'll also cover any person or organization legally responsible for the use of a car, if it's used by you or with your permission. But again, the use has to be for the intended purpose.

You loan your station wagon to a teacher to drive a group of children to the zoo. She and the school are covered by this policy if she actually drives to the zoo, but not if she lets the children off at the zoo and drives to her parents' farm 30 miles away.

By 1978, St. Paul had reduced the number of policy forms in its commercial business package from 366 to 150. Plans now call for most of its commercial insurance policies to be rewritten and reprinted in a more readable format by 1982. The company has appointed a Manager of Forms Simplification, with his own staff and a detailed style manual, to oversee the effort.

The financial field is a particularly promising new area for Plain English efforts. A case in point: Sanford C. Bernstein & Co. is a New York broker-dealer advising clients on investments as well as trading for them. It is subject to regulation under the Federal Securities Exchange Act of 1934 and the Investment Advisers Act of 1940, as well as regulation by the New York Stock Exchange, the National Association of Securities Dealers, the Board of Governors of the Federal Reserve System, and state securities laws in various jurisdictions in which it does business or has clients. These various regulations extend to every aspect of the firm's business, from the content of advertising and soliciting materials to client reports, statements, and the consents that must be obtained before the firm can effect certain types of transaction on clients' behalf. Nonetheless, the firm was able to replace jargon with plain English, while satisfying legal and technical requirements.

Before

It is the express intention of the undersigned to create an estate or account as joint tenants with rights of survivorship and not as tenants in common. In the event of the death of either of the undersigned, the entire interest in the joint account shall be vested in the survivor or survivors on the same terms and conditions as theretofore held, without in any manner releasing the decedent's estate from the liability provided for in the next preceding paragraph.

After

Other signers share your interest equally. If one of you dies, the account will continue and the other people who've signed the agreement will own the entire interest in it.

Plain language scorecard

• Thirty-four states have laws or regulations setting standards for clear language in insurance policies. In 1978, New York became the first state to require that business contracts "primarily for personal, family, or household purposes" be plainly written in everyday language. Since then, New Jersey, Maine, Connecticut, and Hawaii have passed similar laws requiring consumer contracts to be understandable to the public, and a score of other states are considering such laws.

• Hundreds of corporations across the country are simplifying their documents—employee benefit manuals, brokerage account agreements, trusts, the notes to financial statements, customer correspondence, billing statements, and internal communications, ranging from corporate policy manuals to the humble memo.

• President Carter, following up a promise for "regulations in plain English for a change," issued two unprecedented Executive Orders to Federal agencies telling them to simplify paperwork and eliminate gobbledygook from regulations. Agencies that have started to do so include the Environmental Protection Agency and the Department of Health and Human Services (formerly HEW). The Civil Aeronautics Board, where Alfred Kahn started a push for Plain English, is rewriting the notices it requires to be posted at airline counters and printed on the backs of tickets. The Federal Acquisition Regulation Project (FARP), in the Office of Federal Procurement Policy (Department of Management and Budget), offers great potential: FARP is rewriting and recodifying the massive Federal regulations on procurement of goods and services.

• Major law firms, such as Shearman & Sterling in New York, have launched programs to train their young lawyers in clear legal drafting.

• The Internal Revenue Service committed over one million dollars in an all-out effort to simplify the Federal income tax forms. Simplified prototypes were presented to the Congress in November.

• The 1978 Amendments to the Constitution of the State of Hawaii include a provision that, "Insofar as practicable, all governmental writing meant for the public . . . should be plainly worded, avoiding the use of technical terms."

• On April 15, 1980, Gov. Hugh Carey of New York issued an Executive Order directing all state agencies to write their forms and regulations in plain language.

• Last fall, New York City's Department of Consumer Affairs proposed simplified language for more than 40 of its regulations covering deceptive and unconscionable business practices.

• The prestigious Practising Law Institute held a program and published a course book on "Drafting Documents in Plain Language." Another such program is planned for 1981.

• A lobbying group for the movement, Plain Talk, Inc., has been established in Washington, D.C.

• The National Institute for Education, a Federal research agency, has funded the Document Design Project, which is being run by the American Institutes of Research in conjunction with Siegel & Gale and Carnegie-Mellon University. The project's purpose is to study and encourage the use of simplified public documents.

—A.S.

Where obscure terms could not be eliminated, the new forms explain them. The meaning of the term "clearing agents," for instance, is described in this way:

Clearing agents. We can use other broker/dealers as clearing agents to execute transactions, to hold securities in custody on your behalf and to perform other routine procedures in connection with your account. These clearing agents will act at our direction, and they won't have any part in investment decisions.

Especially complex ideas are not only explained but accompanied by illustrative examples:

Buying on margin. In managing your account, we may buy securities on margin. This means that we'll loan you part of the cost of the securities and charge you interest on the loan. By buying on margin, we can purchase more securities on your behalf than we could using only the cash and securities actually available in your account at the time. Because it increases the amount we can invest on your behalf, buying on margin can increase the return on your investment. However, it can also increase the amount of loss on your account.

Example: Assume that the value of A Manufacturing Corp. stock in your portfolio has increased from $5,000 to $10,000. Because of changed market conditions, we decide to sell it and buy Z Industries common stock for your portfolio. If, for instance, the margin requirement is 50%, in theory we could purchase $20,000 worth of Z Industries stock for your account by using the $10,000 proceeds from the sale of your A Manufacturing stock and loaning you the additional $10,000.

In government, the promise—and difficulty—of the Plain English movement is illustrated by the case of the Food Stamp program. That program, begun in 1961, was substantially revised by the Food Stamp Act of 1977. As a result, the forms that had been used by the states were made obsolete. There was a bewildering variety of these forms to begin with. Wisconsin's, for instance, was 37 pages long. To help the states design simple, client-oriented forms reflecting the changes in the law, the Food and Nutrition Service (FNS) of the Department of Agriculture commissioned my firm to come up with new forms to offer to the states as models.

Our approach is illustrated by the new titles we gave our forms:

OLD	NEW
Notice of Expiration	Continuing Your Food Stamps
Notice of Eligibility, Denial or Pending Status	Action Taken on Your Food Stamp Case
Tax Dependency Form	Student Tax Report

Significantly, the new forms ran into a special kind of opposition. First, we discovered that not all state governments agreed that the forms should *help*

Plain Research is needed

The writing of Plain English documents is still a developing field. Researchers have discovered that the reader's comprehension has as much to do with what is inside his or her head as it has to do with what is on the printed page. Diagnostic research can help clarify what information must be explained at greater length.

A study conducted by law professor Jeffrey Davis, described in *Virginia Law Review* 63: No. 6, 1977, provides an excellent example. Davis prepared a simplified installment purchase contract for a refrigerator, showed it to shoppers from various backgrounds, and then asked them questions about its contents. The contract included this definition of default: "I will be in default if I fail to pay an installment on time or if I sell or fail to take proper care of the collateral." Yet most of the readers—even the well educated—failed to understand that they would be in default if they carelessly damaged the refrigerator. They persisted in the common but mistaken belief that they could not be in default if they made all their payments on time.

—A.S.

---

people to use the program, as the law intended. Food Stamp recipients are seen as one measure of a state's poverty, so some officials were reluctant to "encourage" their use by simplifying the paperwork. Some states had actually sought to discourage applicants by using forms that needlessly demanded embarrassing information, such as whether the applicant was a drug user or had a criminal record. Some state officials also felt that the language of the new forms, which included words such as "please," "thank you," and "sincerely," was too nice to people wanting money from the government.

Second, we found that state caseworkers themselves often appeared threatened by new, simplified forms. Some felt that their authority was being undermined. For example, people whose applications for food stamps are turned down or who are cut off are allowed by law to request a fair hearing. We embodied this feature of the law in the forms themselves: letters bearing bad news to Food Stamp clients included a perforated tear-off portion that could be sent back to appeal the decision. Some caseworkers objected heatedly: they felt the tear-offs would cast doubts on their competence and would generate "unnecessary" hearings, increasing their already heavy work load.

After more than two years, the objections seem to have been overcome. Our model forms are used now by about half the states, and one FNS official commented earlier this year: "I've never heard a critical comment about the forms themselves." In fact, as in other simplifications projects, researchers got a strong impression that staff objections had as much to do with the fact the forms were new and unfamiliar as with their style or content.

Earlier in this piece I mentioned that some Plain English efforts have confused simplicity with simplemindedness. A major source of this confusion is the quick-and-dirty "readability formula" that equates clarity with short words and sentences. Here is a rider to a health insurance policy obviously written to such a formula. It is both uncommunicative and insulting to the reader's intelligence.

The Drug Special Endorsement

The DRUG SPECIAL is a small Endorsement. It goes with your [name of company] Contract. It depends on that other Contract to say who's covered. It is added coverage, with big extra benefits.

The DRUG SPECIAL gives you special help in paying for DRUGS . . .

Some attempts at simplification produce a kind of black humor. One life insurance company offers a "simplified" policy requiring that, "The Insured must die while this policy is in force." In a grisly parody that might be called "Dick and Jane Become Underwriters," the policy continues:

The Insured's death may be caused by accidental Bodily Injury. If so the Beneficiary may be paid an additional amount. It will be equal to the Face Amount. Three things must all happen. (1) Death must have been caused by Accidental Bodily Injury. Not by sickness. Not by anything else. . . .

Another misuse of the movement is the way some lawyers are cynically using "simplified" language, notably in apartment leases, to misinform consumers and to mislead them by playing on their ignorance. The nation's first plain language law went into effect in New York State in November 1978. One year later, the State Consumer Protection Board found that most revised lease forms "force tenants to surrender nearly every right they have under law." Rosemary S. Pooler, the board's executive director, referring to the lease prepared by the New York City Bar Association, commented: "It is ironic that tenants were in some respects better off with the 'legalese' of the 1965 lease since the . . . 'plain language' lease seems to have taken almost every opportunity to resolve legal issues in favor of landlords, sometimes at the expense of existing statutory and decisional law."

To illustrate her point, Mrs. Pooler gave, among others, the following examples:

RENT PAYMENT PROVISION

Pre-plain language version (1965):	Plain language version (1978):
"The tenant will pay the rent as herein provided."	"Tenant will pay the rent without any deductions, even if permitted by law."

The new document flies in the face of New York State's Multiple Dwelling Law (Sec. 302–a); Real Property Law (Sec. 235–2); Real Property Actions and Proceedings (Sec. 756 and Article 7–A); and several recent court decisions, all of which give tenants the right to reduce or withhold rent under certain circumstances—for instance, when there are serious building code violations.

ASSIGNMENT AND SUBLETTING

Pre-plain language version (1965):
"(Tenant) will not assign this lease or underlet the leased premises or any part thereof without the landlord's written consent, which landlord agrees not to withhold unreasonably."

Plain language version (1978):
"Tenant shall not assign this lease or enter into a sublease unless it is allowed by a law of the State of New York."

Since 1975, New York's Real Property Law (Sec. 226–b) has given tenants the right to request their landlord's permission to sublet or assign their lease. If the landlord withholds permission unreasonably, he must let the tenant break the lease. The 1965 version is a better statement of existing law than the plain language version, which implies that subletting is allowed by state law only in very special situations.

INSTALLATION OF LOCKS, CHAINS, ETC.

Pre-plain language version (1965):
None.

Plain language version (1978):
"Tenant will not, without landlord's written approval: . . . 3. Put in any locks or chain guards or change any lock-cylinders on the doors of the Apartment."

But New York State's Multiple Dwelling Law (Sec. 51–c) gives tenants of multiple dwellings the right to install additional locks without the landlord's permission, provided that the landlord who requests a key is given one.

Though many lawyers have seen the light, some of them—and their corporate clients—misguidedly try to protect themselves by insisting on too-precise standards for compliance. Such traditionalists do not like the Plain English laws that have been passed in New York, Maine, and Hawaii. Those laws simply require that each document for a residential lease or for a loan, property or services of less than $50,000 for personal, family or household purposes be:

1. Written in a clear and coherent manner using words with common everyday meanings;

2. Appropriately divided and captioned by its various sections.

Ironically, some lawyers and executives who are quick to decry regulatory minutiae in other areas object to this general approach to plain language legislation. They yearn for the false security of the simplistic "readability formula" approach, which ignores less easily quantified elements such as grammar, logic, and organization. They fear that without an exact definition of "clear, coherent, everyday language," compliance will elude them. Lawyers such as Wilbur H. Friedman, chairman of the New York County Lawyers' Association

Safe English

The New York Times (Aug. 31, 1980) reported the first case under the Plain English Law on contracts:

"The New York State Attorney General was puzzled by this sentence.

'The liability of the bank is expressly limited to the exercise of ordinary diligence and care to prevent the opening of the within-mentioned safe deposit box during the within-mentioned term, or any extension or renewal thereof, by any person other than the lessee of his duly authorized representative and failure to exercise such diligence or care shall not be inferable from any alleged loss, absence or disappearance of any of its contents, nor shall the bank be liable for permitting a co-lessee or an attorney in fact of the lessee to have access to and remove the contents of said safe deposit box after the lessee's death or disability and before the bank has written knowledge of such death of disability.'

"Saying, 'I defy anyone, lawyer or lay person, to understand or explain what that means,' Attorney General Robert Abrams sued the Lincoln Savings Bank in New York City . . . demanding that it simplify a customer agreement on safe-deposit boxes.

"The case is settled . . . The former 121-word sentence now says: 'Our liability with respect to property deposited in the box is limited to ordinary care by our employees in the performance of their duties in preventing the opening of the box during the term of the lease by anyone other than you, persons authorized by you or persons authorized by the law.' "

Special Committee on Consumer Agreements, direfully predicted that New York's law would create "upheaval" among businesses and an "absolutely staggering" burden on the courts.

But these fears were unfounded: there has been no flood of lawsuits. The arguments of the technically minded still have their appeal, however, as demonstrated by Connecticut's plain language law. It lists two alternate sets of criteria for plain language, with nine and eleven different tests respectively, including these:

1. The average number of words per sentence is less than twenty-two; and

2. No sentence in the contract exceeds fifty words; and

3. The average number of words per paragraph is less than seventy-five; and

4. No paragraph in the contract exceeds one hundred fifty words; and

5. The average number of syllables per word is less than 1.55; and . . .

Cookbook detail like this only guarantees that the spirit of plain language legislation will be lost in attempts to follow it to the letter: the clarity of a sentence becomes less important than seeing that it has "less than 1.55" syllables per word. And what would happen to interstate commerce if one of Connecticut's neighbors were to require less than 20 words per sentence instead of 22, or 160 words per paragraph instead of 150?

Keeping plain language laws simple, like the law in New York State, is essential. The courts should be left free to judge particular cases according to a general standard, as they do now with the concept of the "reasonable man." Lawyers who hang back, in favor of the traditional approach to legal language, should remember the traditional result: as Harold Laski put it, in every revolution the lawyers lead the way to the guillotine.

Non-discriminatory writing

WILLIAM C. PAXSON

William C. Paxson is an editor and writer in Sacramento, California. His articles have covered a wide variety of topics, including air pollution, electronics, travel, and school administration.

*L*anguage shapes thought. If language didn't shape thought, words like *propaganda* and *publicity* wouldn't be in dictionaries. If language didn't shape thought, politicians wouldn't be elected. If language didn't shape thought, Hitler would never have gotten out of the beer halls.

In shaping thought, language has the power to reinforce and perpetuate ethnic, racial, or sexual discrimination. Language used in this manner works against attempts to achieve equal opportunity in education, housing, and jobs; and equal opportunity is a national philosophy supported by law.

Discrimination in writing can be traced in part to the myths and stereotypes that have become associated with certain groups. This association has produced a form of folklore in which Italians are swarthy, Chinese are smiling and serene, blacks are athletic and shuffling, Mexicans are lazy, Spaniards are passionate, and native American Indians are proud and stoic.

Another form of folklore leads to discrimination on the basis of sex. This form stresses the concepts of male superiority and female inferiority. In this particular folklore, all positions of leadership are held by men, and all low-status occupations and activities are performed by women.

It would take a book larger than this one to trace the source of such attitudes. Still, we cannot deny their existence, and they do find their way into writing. A personnel report that evaluates Mary Smith as "a well-groomed black clerk" implies that other blacks are not well-groomed. An ad for a manager that reads "he is required to monitor production and he must act decisively" carries with it the message that only a man can handle the job.

In reality, attitudes such as these deny the fact that all sorts of personal attributes can be found in all groups and individuals. Therefore—don't label. All Asians are not inscrutable. All Anglo-Saxons are not poised and prim. All Scots are not thrifty. Many elderly people detest being called *senior citizens*. There is no such person as a *typical teenager*. All women are not weak and flighty, and all men are not strong, logical thinkers. A man's wife is not his *little woman*, and females in college are not *coeds*; they are students.

Another cause of discrimination in writing is found in the noun *man* and the pronouns *he*, *him*, and *his*. These traditionally have been used to apply to both men and women. Thus, we see statements like "Man works to protect nature"; "The manager is a busy person, and he needs the essential details of a report quickly"; and "An employee is encouraged to express his opinions about morale."

Granted, the writers of sentences such as these may not mean to exclude women. Still, the words literally say "men only," and the writing can be judged as discriminatory.

What is needed is a neutral noun or pronoun that applies equally to both sexes. The English language does not have such a word, but other alternatives are to:

1. Use the male pronouns *he*, *him*, and *his* when discussing a specific man only and not to lump everybody together regardless of sex.

2. When discussing people without regard to sex, use the genderless *one*, words such as *person* or *individual*, or the second person *you*, *your*, or *yours*. Thus, "his position in the organization" becomes "one's position"; "this person's position"; or "your position."

You can also neuter prose with *the* in certain situations. As an example, instead of writing of an engineer using "his test meter" write "the test meter."

In some instances, plurals work out. Thus "a man's opportunity is unlimited" becomes "their opportunity is unlimited" or "these people have unlimited opportunity."

Still another method is to write *he* or *she* instead of just *he* when both sexes are being referred to. When this is done, "he must have the appropriate technical ability" becomes "he or she must have the appropriate technical ability." Alternating between *he* and *she* is not limited to single phrases but can be done from sentence to sentence, paragraph to paragraph, or chapter to chapter.

But using *he* or *she* consistently in a short publication or alternating between genders in the parts of a longer publication poses problems a writer must be aware of. First, the frequent use of *or* is a weakness in style. Second, the frequent used of *he* or *she* or *his* or *her* bounces the reader back and forth, and the bouncing is tiring. If you can use some of the other methods suggested here, your writing will be stronger and show more polish.

Also, remember that language is first a spoken tool and then a written one. What sounds good usually reads that way. Therefore, if we are going to preserve the strength of our language we will have to avoid unpronounceable combinations such as *he/she* or *s/he*. How does one pronounce those crisply? Try it out loud: "he slant she," "he or she," "he and she"; or "ess slant he," "she he"? Which is right? What is meant? What does it *sound like*?

3. To further reduce discrimination in writing, use nonsexual titles and references:

INSTEAD OF	WRITE
man-made	synthetic, artificial
manpower	workers, workforce
man-hours	working hours, staff hours
gentlemen's agreement	informal agreement
you and your wife	you and your spouse
chairman	chair
businessman	executive, entrepreneur
foreman	supervisor
salesman	sales agent

4. When using names and courtesy titles, refer to men and women equally:

INSTEAD OF	WRITE
Roger Smith and Miss Jones	Roger Smith and Sarah Jones
Mr. Smith and Sarah Jones	Mr. Smith and Miss Jones
Roger and Jones	Roger and Sarah

Along these same lines, if you are writing to a woman and if you don't know if she is a *Miss* or *Mrs.*, use *Ms. Ms.* is generally accepted as being the female equivalent of *Mr.* The rule is to address women by the same standards as one addresses men.

5. Lastly, don't get carried away. *Host* and *hostess* are usable; *adulterer* and *adulteress* are specific and correct; and *leading lady* and *leading man* are bonafide titles.

Also, don't write *Black* unless you are willing to give whites equal upper-case time.

And if you are writing to an addressee whose gender is unknown, your options are: *To whom it may concern, Dear Sales Director, Dear XYZ Company, Dear Sir or Madam,* or perhaps no salutation at all. In any event, a term like *Dear Gentlepeople* is strictly a last resort. A writer in an organization where I worked did that once and got this reply: "Dear Sir. In an oil field, there are no gentlepeople."

3

Business and technical correspondence

How to write a business letter

MALCOLM FORBES

Malcolm Forbes is President and Editor-in-Chief of *Forbes Magazine*.

A good business letter can get you a job interview.

Get you off the hook.

Or get you money.

It's totally asinine to blow your chances of getting *whatever* you want—with a business letter that turns people off instead of turning them on.

The best place to learn to write is in school. If you're still there, pick your teachers' brains.

If not, big deal. I learned to ride a motorcycle at 50 and fly balloons at 52. It's never too late to learn.

Over 10,000 business letters come across my desk every year. They seem to fall into three categories: stultifying if not stupid, mundane (most of them), and first rate (rare). Here's the approach I've found that separates the winners from the losers (most of it's just good common sense)—it starts *before* you write your letter:

Know what you want

If you don't, write it down—in one sentence. "I want to get an interview within the next two weeks." That simple. List the major points you want to get across—it'll keep you on course.

If you're *answering* a letter, check the points that need answering and keep the letter in front of you while you write. This way you won't forget anything—*that* would cause another round of letters.

And for goodness' sake, answer promptly if you're going to answer at all. Don't sit on a letter—*that* invites the person on the other end to sit on whatever you want from *him*.

Plunge right in

Call him by name—not "Dear Sir, Madam, or Ms." "Dear Mr. Chrisanthopou-los"—and be sure to spell it right. That'll get him (thus, you) off to a good start.

(Usually, you can get his name just by phoning his company—or from a business directory in your nearest library.)

Tell what your letter is about in the first paragraph. One or two sentences. Don't keep your reader guessing or he might file your letter away—even before he finishes it.

In the round file.

If you're answering a letter, refer to the date it was written. So the reader won't waste time hunting for it.

People who read business letters are as human as thee and me. Reading a letter shouldn't be a chore—*reward* the reader for the time he gives you.

Write so he'll enjoy it

Write the entire letter from his point of view—what's in it for *him*? Beat him to the draw—surprise him by answering the questions and objections he might have.

Be positive—he'll be more receptive to what you have to say.

Be nice. Contrary to the cliché, genuinely nice guys most often finish first or very near it. I admit it's not easy when you've got a gripe. To be agreeable while disagreeing—that's an art.

Be natural—write the way you talk. Imagine him sitting in front of you— what would you *say* to him?

Business jargon too often is cold, stiff, unnatural.

Suppose I came up to you and said, "I acknowledge receipt of your letter and I beg to thank you." You'd think, "Huh? You're putting me on."

The acid test—read your letter *out loud* when you're done. You might get a shock—but you'll know for sure if it sounds natural.

Don't be cute or flippant. The reader won't take you seriously. This doesn't mean you've got to be dull. You prefer your letter to knock 'em dead rather than bore 'em to death.

Three points to remember:

Have a sense of humor. That's refreshing *anywhere*—a nice surprise in a business letter.

Be specific. If I tell you there's a new fuel that could save gasoline, you might not believe me. But suppose I tell you this:

"Gasohol"—10% alcohol, 90% gasoline—works as well as straight gaso-line. Since you can make alcohol from grain or corn stalks, wood or wood waste, coal—even garbage, it's worth some real follow-through.

Now you've got something to sink your teeth into.

Lean heavier on nouns and verbs, lighter on adjectives. Use the active voice instead of the passive. Your writing will have more guts.

Which of these is stronger? Active voice: "I kicked out my money manager." Or, passive voice: "My money manager was kicked out by me." (By the way, neither is true. My son, Malcolm Jr., manages most Forbes money—he's a brilliant moneyman.)

Give it the best you've got

When you don't want something enough to make *the* effort, making *an* effort is a waste.

Make your letter look appetizing—or you'll strike out before you even get to bat. Type it—on good-quality 8½″ × 11″ stationery. Keep it neat. And use paragraphing that makes it easier to read.

Keep your letter short—to one page, if possible. Keep your paragraphs short. After all, who's going to benefit if your letter is quick and easy to read?
You.

For emphasis, underline important words. And sometimes indent sentences as well as paragraphs.

Like this. See how well it works? (But save it for something special.)

Make it perfect. No typos, no misspellings, no factual errors. If you're sloppy and let mistakes slip by, the person reading your letter will think you don't know better or don't care. Do you?

Be crystal clear. You won't get what you're after if your reader doesn't get the message.

Use good English. If you're still in school, take all the English and writing courses you can. The way you write and speak can really help—or *hurt*.

If you're not in school (even if you are), get the little 71-page gem by Strunk & White, *Elements of Style*. It's in paperback. It's fun to read and loaded with tips on good English and good writing.

Don't put on airs. Pretense invariably impresses only the pretender.

Don't exaggerate. Even once. Your reader will suspect everything else you write.

Distinguish opinions from facts. Your opinions may be the best in the world. But they're not gospel. You owe it to your reader to let him know which is which. He'll appreciate it and he'll admire you. The dumbest people I know are those who Know It All.

Be honest. It'll get you further in the long run. If you're not, you won't rest easy until you're found out. (The latter, not speaking from experience.)

Edit ruthlessly. Somebody has said that words are a lot like inflated money—the more of them that you use, the less each one of them is worth.

~~Right on.~~ Go through your entire letter ~~just~~ as many times as it takes. ~~Search out and~~ Annihilate all unnecessary words, ~~and~~ sentences—even ~~entire~~ *para-graphs.*

Sum it up and get out

The last paragraph should tell the reader exactly what you want *him* to do—or what *you're* going to do. Short and sweet. "May I have an appointment? Next Monday, the 16th, I'll call your secretary to see when it'll be most convenient for you."

Close with something simple like, "Sincerely." And for heaven's sake sign legibly. The biggest ego trip I know is a completely illegible signature.

Good luck.

I hope you get what you're after.

Sincerely,

Making your correspondence get results

DAVID V. LEWIS

David V. Lewis is an in-house consultant with Western Company of North America in Fort Worth, Texas, where he is involved in sales and management training.

*I*f you turn out an average of five letters a day, you'll produce nearly twice as many words during the year as the typical professional writer.

These letters reflect *your* attitude—and obviously your organization's—toward customers. Realizing this, many progressive organizations train key people in the art of writing readable, results-getting letters. For example, the New York Life Insurance Company produces more than a million letters a year and uses its correspondence as a public relations tool.

"Anyone who writes a letter for New York Life holds a key position in our organization," says Nathan Kelne, vice president of the company. "By the letters our people write, they help determine how the public feels toward our company, and toward the life insurance business as a whole. Since they are instrumental in shaping the personality of the company, they are in a very real sense public relations writers."

For example, before New York Life launched its companywide letter-writing courses, here's how one of its executives tried to explain to a beneficiary the way the death claim was to be paid on a $3,000 policy:

> The monthly income per $1,000 under option 10 years certain is $7.93 per $1,000 total face amount of insurance, which total is $3,000.

Fortunately, the beneficiary was able to cut through the jargon. He replied:

> As I understand your letter, you seem to be saying that I should receive a monthly check for the amount of $23.79 for at least 10 years.

But another policyholder's response to a similarly bewildering letter is more typical:

> Please tell me what you want me to do, and I'll be glad to do it.

Some universally proven principles that can help you sell yourself and your company to the public will now be examined.

Write for him, not to him

General Foods is an organization that believes strongly in creating a favorable image through its correspondence. The public relations people came across this letter signed by a marketing executive and ready for mailing:

> Dear Sir:
> Enclosed please find a questionnaire which we are sending to all our retail contacts in this state.
> Will you please answer this as soon as possible? It's very important that we have an immediate reply. We're delaying final plans for our retail sales program in this area until we get answers to the questionnaire.
> With thanks in advance, we are,
>
> Gratefully yours, . . .

The letter is clear and to the point; it *does* communicate readily. But there's a major flaw. It points out benefits the company will receive instead of suggesting how the program will help the recipient. The letter is writer- rather than reader-oriented.

Psychologists say that each of us is basically interested in himself or herself. We want to know "what's in it for me?" Once you routinely approach letter writing from this point of view, you'll find yourself telling your readers in specific terms how the letter will benefit them.

The questionnaire letter was rewritten this way.

> Dear Sir:
> Enclosed is a questionnaire on our proposed retail sales program. Your advance opinion on how this new plan would help you and others will be very useful in our evaluation of the program.
> If you will fill out and return this questionnaire as soon as possible, we can let you and other retail contacts in this area know promptly of changes in the present program. Thank you.
>
> Very truly yours, . . .

The revised version asks for your advance opinion, suggests how quick action might help you and *others,* and promises to let *you* know promptly of changes. There's also another improvement. The tired and outdated "With thanks in advance" has been replaced by a modern "Thank you."

The best way to persuade your readers to your point of view is to show them that it will be worth their while to do so. This rule holds true for just about every successful letter, sermon, sales presentation, or advertisement. A good sales letter, for example, is much like a good ad in that it attempts to dramatize benefits to the reader.

What does it take to get—and hold—a reader's attention? The most powerful letters appeal to basic needs and emotions rather than to purely logical reasons. Mainly, management people want to know how to save time, money, and effort. Show them how your product or service can help them do one or more of these things, and you're likely to get their attention.

With effort, almost any letter can be oriented to the reader, even the usually hard-to-write collection letter. Here's such a specimen, written almost entirely from the writer's point of view:

> Dear Sir:
> Our records show that you are three months delinquent in payment of your bill for $37.50.
> Perhaps this is an oversight on your part. Otherwise we cannot understand why you have not taken care of this obligation.
> Very truly yours, . . .

Keeping the reader's self-esteem in mind, the letter could have been rewritten this way:

> Dear Sir:
> We know you'll want to take care of your small past-due account for $37.50. This will help you to maintain the fine credit you have built up with us over the years.
>
> Sincerely, . . .

To help develop the "you" attitude, put yourself in your reader's shoes, then write from his or her point of view. Once you've developed this attitude, you'll automatically start telling your readers what's in it for them. Instead of saying, "I wish to thank you," you'll write, "Thank you." Instead of writing, "We'd like to have your business," you'll write, "You'll find our service can help your business in many ways."

Personalize your letters (there's power in pronouns)

Some years ago, many would have considered this letter to have been perfectly proper and very effective:

> Gentlemen:
> Enclosed herewith are the subject documents which were requested in yours of the 10th. The documents will be duly reviewed and an opinion rendered as to their relevancy in the involved litigation.
> Very truly yours, . . .

The letter *was* all right—back in the horse-and-buggy days! Phrases like "enclosed herwith," "duly received," and "involved litigation" would have marked the writer as learned. But the executive who makes a habit of writing like that today is generally regarded as an anachronism.

Current usage calls for clear, to-the-point letters, written mostly in conversational language. Like good conversation, your letters should generally be friendly, filled with personal references, and almost always informal in tone and language. Here's how the horse-and-buggy letter might have been rewritten by the modern executive:

> Dear Sam:
> Here are the documents you asked for. I'll look them over and let you know if we can use them in our lawsuit.
>
> Cordially, . . .

When experts write a letter, even a form letter, they generally try to make it sound as personal as possible. Almost always, it contains a sprinkling of personal pronouns. In orienting your letter to the reader, fill it with "you," "your," and "yours." Use "I" and "me" sparingly.

Here's a case in point, a letter sent out by a mortgage company (emphasis added by the author):

> Dear Mr. Jones:
> *We* want to thank you for your query about *our* new mortgage insurance plan.
> *We* are enclosing a pamphlet which outlines benefits of *our* new policy and gives testimonials from some of *our* policyholders.
> *We* would like very much to enroll you within the next 30 days, since *we* are offering a special low premium rate as *our* introductory offer.
>
> Very truly yours, . . .

The repeated use of *we* and *our* clearly shows the letter is written with the mortgage company's interests at heart. But notice how substituting *you* and *yours* orients the letter to the reader's interest. (Emphasis has again been added by the author.)

> Dear Mr. Jones:
> Thank *you* for *your* recent inquiry about our new mortgage insurance plan.
> *You'll* find that the enclosed pamphlet outlines benefits of the policy in detail. *You'll* also probably be interested in the testimonials furnished by some policyholders.
> If *you* sign up within the next 30 days, *you'll* be able to take advantage of special premiums we're offering on an introductory basis. Thank you.
>
> Cordially, . . .

The word is out now in enlightened business circles. Companies are telling their executives to regard every letter as a personal contact: to write your own way, to develop your own style (within bounds), to make your letter distinctively *you*.

Mastering tone (your personality in print)

Writing a "rejection letter" that leaves the recipient's self-esteem intact and preserves his or her goodwill is a difficult task. It requires tact, diplomacy, and empathy—all of which must be effected through appropriate *tone*. For example, here's the way one banker turned down a builder, a long-time customer:

> Dear Mr. Jones:
>   We regret to inform you that your request for an additional loan in the amount of $250,000 must be rejected. It was the judgment of the loan committee that, with your present commitment, such a loan would present too much of a risk for us.
>   Naturally, we look forward to doing business with you on your existing commitment.
>
>   Sincerely, . . .

What's wrong: Tone, mainly—the *way* the rejection is phrased. It deflates the reader's ego and makes it virtually certain the builder will do his banking elsewhere in the future.

True, the rejection was "justified." But why not let the reader down more gently and leave the door open for future business? For example:

> Dear Mr. Jones:
>   One of the most distasteful tasks we have is to turn down a loan application, particularly when it is from a regular and respected customer like you.
>   We know you'll be disappointed, and so are we. We share your excitement about your plans to expand the Roseland Project, and hope that circumstances will later warrant our working with you on this and other projects.
>   However, the current economic outlook, combined with your delinquent status on your existing loan, makes your new loan application a questionable venture for us at the present time.
>   We will be glad to work with you in any way we can to resolve your current financial problems, and we look forward to helping you meet your future financial commitments.
>
>   Sincerely, . . .

The two letters say the same thing, but in vastly different ways. Suffice it to say that tone is as important to a letter as good muscle tone is to an athlete.

A negatively phrased letter can have the same effect as an abrasive personality. As one New York Life executive put it:

> By its very nature, the "no" letter is a turndown and leaves the reader unsatisfied. Yet there are ways to soften the blow, and one of the best is simple candor. So, if you must say "no," state the reason first: "Currently there are no vacancies in that department. Therefore, I cannot offer you a position with the company at this time."
>
> A "no" response often requires positive alternatives if they are appropriate (for example, "Although, for the reasons mentioned, I cannot do as you ask, may I suggest that . . . "). You have to be more diplomatic and more sensitive if you're to have any chance of salvaging your reader's goodwill or ego.
>
> And that really is your goal in an effective "no" letter—to tell your readers something they don't want to hear in a way that compels understanding and, ideally, acceptance.

Unfortunately, tone requirements vary not only from person to person, but from one situation to another. For example, you would ordinarily use fast-paced, persuasive prose for a sales-promotion letter. In writing to a highly regarded law firm, you might use a slightly more "dignified" approach. If you're writing for a service-type organization, you might use a middle-of-the-road approach, as the government does.

Government letter writers are told to strive for a tone of "simple dignity," to make letters brief and to the point, and to avoid gobbledygook. "Don't act as if you're the only game in town," they're told. "On the other hand, don't bow and scrape just because you're a service organization."

Poor letters start with poor thinking on the part of the writer. Negative thoughts lead to negative words—and before you know it, there goes the old ball game. Here's a letter from a manager of a department store to a customer who complained that an appliance she had recently bought didn't work. (Negative words have been italicized by the author.)

Dear Sir:

I am in receipt of your letter in which you *state* that the hair dryer you purchased from us recently *failed to meet* the warranty requirements.

You *claim* that the dryer *failed* to do the things *you say* our salesman promised it would do.

Possibly *you misunderstood* the salesman's presentation. Or perhaps *you failed to follow instructions* properly. We *positively* know of no other customer who has made a similar complaint about the dryer. The feeling is that it will do all that is stated *if properly used*.

However, we are willing to make some concessions for the *alleged* faulty part. We will allow you to return it; however, *we cannot* do so *until* you sign the enclosed card and return it to us.

Very truly yours, . . .

The tone is unmistakably negative. "State" and "failed to meet" in the opening paragraph imply there's some doubt that the claim is valid. In the second paragraph, "you say" suggests the salesman didn't make the statement at all.

Such phrases as "you misunderstood," "you failed to follow instructions," and "if properly used" tell you in so many words that you're not too bright. And to wrap it up, the writer uses the negative "we cannot . . . until" instead of the positive "we will . . . as soon as."

Most letters aren't this negative, of course. But it doesn't take much to offend. Any of these negative words or phrases, in themselves, could have spoiled the tone of an otherwise effective letter.

If you've been guilty of taking a negative approach, study the following examples. In each case, the negative thought (emphasis added by the author) has been converted into a positive one.

Negative	Since you *failed* to say what size you wanted, we cannot send you the shirts.
Positive	You'll receive the shirts within two or three days after you send us your size on the enclosed form.
Negative	We *cannot* pay this bill in one lump sum as you requested.
Positive	We can clear up the balance in six months by paying you in monthly installments of $20.
Negative	We're sorry we *cannot* offer you billboard space for $200.
Positive	We can offer you excellent billboard space for $300.
Negative	We are *not* open on Saturday.
Positive	We are open from 8 A.M. to 8 P.M. daily, except Saturday and Sunday.

Negative words aren't the only cause of poor letter tone. Many letters are made more or less "neutral" by mechanical, impersonal, or discouraging language. Here's an example of each fault with a preferred alternative:

Impersonal	Many new names are being added to our list of customers. It is always a pleasure to welcome our new friends.
Personal	It's a pleasure to welcome you as our customer, Mr. Jones. We'll make every effort to serve you well.
Mechanical	This will acknowledge yours of the 10th requesting a copy of our company's annual report. A copy is enclosed herewith.
Friendly	Thanks for requesting a copy of our annual report, which is enclosed. We hope you will find it helpful.
Discouraging	Since we have a shortage of personnel at this time, we won't be able to process your order until the end of the month.
Encouraging	We should have more help shortly, which will enable us to get to your order by the end of the month.

Ideally, the tone of your letter should reflect the same ease in conversation that you enjoy when talking about your favorite hobby, business, or pastime.

How to write (more) the way you talk

This letter was sent out by a large department store. Imagine yourself as the recipient.

Dear Sir:

We are in receipt of yours of July 10 and contents have been duly noted.

As per your request, we are forwarding herewith copies of our new fall brochure. Thanking you in advance for any business you will be so gracious as to do with us,

Yours truly, . . .

Conversational? Of course not. Who uses such language as "We are in receipt of . . . ," "duly noted . . . ," and "as per your request"? Practically no one. People just don't talk that way. Most of these phrases went out with the Model-T—or should have.

Face to face, the writer probably would have said, "Here's the fall brochure you requested. Let me know if we can serve you." It says the same thing, in less space, and without all the fuss.

Some business and professional people still feel it isn't quite proper to write conversationally, mainly because they have seen so much stilted business writing. But many progressive companies are telling their people to communicate in plain English, using only those technical terms that are absolutely necessary. As one executive of a major company said, "The best letters are more than just stand-ins for personal contact. They bridge any distance by the friendly way they have of talking things over person to person." And this from a U.S. Navy bulletin: "At best, writing is a poor substitute for talking. But the closer our writing comes to conversation, the better our exchange of ideas will be."

The consensus clearly is that informal, natural business writing is in; stilted business writing is *out*.

One word of caution about writing (more) the way you talk. Since World War II, readability experts have urged business people to "write more the way they talk" or "write as they talk."

Detractors have soft-pedaled the idea, claiming that most conversation is rambling, often incoherent, and frequently a bit too earthy. These are valid objections, but they miss the point. You're not being asked to write *exactly* the way you talk. Rather, you're being asked to bridge the gap between the spoken and the written word—to narrow the difference in the way you would *give* an order verbally and the way you would *write* that same order in a memo.

Naturally, writing requires more restraint than speech, and the writer must normally use fewer words. But you can do these things and still capture the tone and cadence of spoken English.

The first step in making your correspondence more conversational is to rid your vocabulary of worn-out business phrases. Here are some of the more flagrant offenders. They were stylish once, but they've done their duty and need to be honorably discharged.

OLD HAT	CONVERSATIONAL
At a later date	Later
If this should prove to be the case	If this is the case
This will acknowledge receipt of	Thank you for
Attached herewith please find	Here is; Enclosed is
We shall advise you accordingly	We'll let you know
Due to the fact that	Because
With regard to	About
Please notify the writer as to	Please let me know
Enclosed please find a stamped envelope	I've enclosed a stamped envelope
We are submitting herewith a duplicate copy	Here is a copy
In compliance with your request	Here is
We are submitting herewith our check in the amount of $75	Here is our check for $75
We beg to advise (acknowledge) that	[Begging is unnecessary]
The information will be duly recorded	We'll record the information
The subject typewriter	This typewriter
In compliance with your request	As you requested
We will ascertain the facts and advise accordingly	We'll let you know
The writer wishes to state	[Just say it]

This list is far from complete but you get the idea. Once you're mentally geared to writing more conversationally, you'll detect many other clichés. Try to eliminate them.

Contractions will also make your writing more conversational. They play a part—a very large part—in almost everyone's everyday conversation. Even your most learned associate doesn't say, "We shall endeavor to be there at eight o'clock"; he's more likely to say, "We'll try to be there at eight o'clock." Instead of saying, "I am going to the ball game," he'll probably say, "I'm going to the ball game." It takes less effort to say, "You needn't bother to call," than to say, "You need not bother to call." Contractions tend to make the spoken words flow more smoothly. That's why they make writing appear more natural.

Probably the most common contractions are here's, there's, where's, what's, let's, haven't, hasn't, hadn't, won't, wouldn't, can't, couldn't, mustn't, don't, doesn't, didn't, aren't, isn't, and weren't. Then there are the pronoun contractions: I'll, I'm, I'd, I've, he's, he'll, he'd, and so forth.

Using these and other contractions when they facilitate flow of words will do much to give your writing a quality of spontaneity and warmth. But they must be used with discretion. Using too many contractions can sometimes make your writing too informal. And used in the "wrong" place, they might not "sound right." Indeed, the key is whether the contraction sounds right when the sentence is read.

Take this section from the Gettysburg Address: "But in a larger sense, we cannot dedicate, we cannot consecrate, we cannot hallow this ground." The passage would undoubtedly have lost its historic tone if *can't* had been used instead of *cannot*.

On the other hand, the advertising slogan "We'd rather fight than switch" would lose some of its punch if phrased, "We would rather fight than switch." Appropriate usage depends on how the contraction makes the sentence sound when it's read aloud.

Next time you get ready to write a letter, ask yourself, "How would I say this if I were talking to the person?" Then go ahead and write in that vein.

Letters
that
sell

THE ROYAL BANK OF CANADA

*E*veryone writes letters that sell, and every letter has as its purpose the selling of something: goods, services, ideas or thoughts.

Someone may say that a family letter has no such purpose, but consider this: a letter telling about the children seeks to promote a favourable impression of their welfare and happiness; a letter telling about illness is designed to gain sympathy; the letter that says nothing but "I hope you are well" is selling the idea "I am thinking of you".

Family letters are usually rambling letters. They would be improved both in their readability and their informativeness if they adopted some of the principles that are used to sell goods and services. Business building letters, on the other hand, could with advantage incorporate some of the friendly informality of family letters.

Salesmanship of any kind is basically a person moving goods by persuading another person that he needs them, or winning that person's support or approval of an idea or a plan.

Some non-commercial type sales letters are those that champion good causes, such as community welfare or health standards or national unity. They seek to influence the thinking of individuals or groups.

It is not a simple task to compose a letter designed to sell. Like any other product of value, it calls for craftsmanship. There are techniques to be learned, techniques of conveying ideas, propositions, conclusions or advice appealingly and purposefully.

In the beginning

In creating a letter to sell something we need to begin by thinking about the person to whom we are writing. A lawyer studies his opponent's case just as sharply as his client's; the manager of a baseball or hockey team analyses the qualities, good and bad, of members of the opposing team.

The writer must anticipate and answer in his letter questions that will occur to the reader: What is this about? How does it concern me? How can you prove it? What do you want me to do? Should I do it?

People buy goods or services because these will give them a new benefit or will extend or protect a benefit they already have, so the writer needs to translate what he offers into owner benefits.

The proffered benefits must be accessible and adapted to the reader's position, environment and needs. No letter is likely to sell sun-bonnets to people who live beyond the Arctic Circle or baby carriages to bachelors. We may classify a potential customer as a man, woman, company or institution that will have use for a product or service, has sufficient money to pay for it, and in whom a desire for possession may be created.

The reader's interest: that is the guiding star in sales letter writing. See his interests, his angle, and accommodate your stance to them. A simple precaution against sending a letter to the wrong person is to ask yourself what use you would have for the commodity if you were in the reader's place.

It is a good rule to spend more time thinking about the reader than about what you have to say. Otherwise you may become wrapped up in the virtues of your product so that you forget that the decision to buy rests with your prospect.

The self-interest of the person to whom you write is a major factor to consider in successful sales communication. When you remember it you give the impression that you have singled out this reader as being an important individual, and that is an excellent introduction.

It is not to be expected that the writer of letters that sell will know every person to whom he writes, but he must know certain facts: approximate income and age, occupational level, his business, and things like that. Then he is able to slant his sales points accurately toward the reader's needs, interests and purchasing power.

Know your product

The reader's attention should be attracted to the product or service, not to the grand style or picturesque phraseology of your letter. When you catch a person's attention you are focusing his consciousness on something. Concentrate on your commodity. The best magnet to draw and hold attention is what you say about the product, showing it to be useful and the means of fulfilling a desire.

It is no small accomplishment to analyse and marshal into order the facts about a product so as to win the thoughtful consideration of a person who has plenty of other things on his mind.

In purchasing almost any sort of commodity the buyer has a choice between what you are offering and what others are selling. Your sales job is to show the superiority of your product. Tell why what you offer is necessary or desirable, what it will accomplish in your reader's business, and how it can be fitted into his present layout and his plans. Do not content yourself with telling about the article as it sits on display: picture it in use in the reader's home or factory.

Your letter needs to convey the assurance that you are telling the truth about your goods. It is not a sensational offer that makes a letter convincing, but the feeling that the reader can depend upon what is said. He should feel assured that he will be buying what he thinks he is buying. Customer dissatisfaction caused by misleading sales talk can cause shock waves that affect the whole selling organization.

Let your personality show

Make your letter sound friendly and human: put your personality on paper. Your letter is you speaking. Some of the features in your personality that you can display are: friendliness, knowledge, keen-mindedness, trustworthiness and interest in the prospect's welfare.

What you have in your mind about the good quality, appearance and usefulness of your product has to be communicated to your reader so as to arouse his interest, create a desire to possess, and induce him to buy.

Communication is not the easiest thing in the world to attain in writing, in art, or in music. Dr. Rollo May wrote in *Man's Search for Himself:* "We find in modern art and modern music a language which does not communicate. If most people, even intelligent ones, look at modern art without knowing the esoteric key, they can understand practically nothing."

It is not enough to write something so that it can be read. The degree to which communication occurs depends upon the degree to which the words represent the same thing for the reader as they do for the writer.

The recipient of a letter that is not clear is likely to blame its opacity on the lack of intelligence of the writer.

The art of composing sales letters is not one to be mastered by minds in which there is only a meagre store of knowledge and memories.

The art consists in having many mental references and associating them with new thoughts. Consider a poem. Its theme will likely have arisen from a single event, but the images used in its construction will have been drawn from the total life experience of the poet.

Put some flavour into your letters so that they taste good. Your letter will not be like anyone else's. That is a virtue, just as being an individual is a virtue in conversation. Who wishes to be a carbon copy of a textbook letter or to parrot phrases that other people use?

Practise talking on paper as if you were on the telephone. First write down the imagined questions asked by the person on the other end of the line and then your answers, given in simple, direct and pleasing words. To humanize your letters in this way with the natural idiom of conversation does not mean that you use cheap slang or clever verbal stunting.

Show some style

The style in which you write is not a casual feature of your letter. It is vital to your reader's understanding of what you are saying to him. It is not your job to please the reader's sense of the aesthetic, but to tell and explain plainly what is necessary to introduce your goods or your idea to his favourable attention. This may be done in a way that has grace and comeliness.

Never "talk down" to a reader. Make him feel that he knows a great deal, but here is something he may have missed. There is a big difference, when trying to build business, between making a suggestion and preaching a sermon.

It is highly important in writing a letter to sell something that it should be appropriate. Whatever your writing style may be, it will fit the occasion if it gives this particular correspondent information that will be useful to him, conveys to him a feeling of your interest in him and his business, and assures him of your goodwill.

Besides being grammatically correct, language should be suitable. At one extreme of unsuitability is the language that is too pompous for its load, and at the other is the language of the street which belittles the receiver's intellectual level.

Your words should be the most expressive for their purpose that the language affords, unobstructed by specialty jargon, and your sentences should be shaken free of adjectives—the most tempting of forbidden fruit to a person describing something.

Properly chosen words will convey your appreciation of the addressee as a person, and such friendliness is contagious. Some people are afraid to be friendly in their letters. They fear they will be thought of as "phonies" who have disguised themselves as Santa Claus for the occasion. Being friendly and showing it should not raise this scarecrow. It would be a grave mistake, indeed, for any of us to indulge in flowery language foreign to our natural talk; but it is no mistake at all to incorporate in our letters the warm, personal language that comes naturally to us in person-to-person social contacts.

Letter writing invites us to use the same etiquette as we use in courteous conversation. We look at the person with whom we are talking, converse on his level of understanding, speak gently, and discuss matters he considers important or interesting.

What the reader of your letter will notice is not its normal courtesy, but the extra touch that demonstrates care and understanding, a genuine interest in the reader's wants, a wish to do what is best for him, and the knowledge you show of how it can be done.

Everyone who writes a letter has a moral as well as a business reason to be intelligible. He is placing his reader under an obligation to spend time reading the letter, and to waste that time is to intrude upon his life plan.

There is an eloquence of the written as well as of the spoken word. It consists in adapting a statement to the receptive system of the reader so that he will have maximum help against confusion, against mistaking what is incidental from what is fundamental. A familiar device to use in this effort is to relate the new commodity you are offering to something that is familiar.

Use suitable formulas

There are formulas you may wish to make use of. Your letter must conform in some respects to what letters are expected to be. This does not mean pouring all letters into the same mould. Within the accepted pattern you are free to develop your talent for expression.

Skill is needed in the use of formulas. A form letter reveals itself to the reader and gets short shrift. It is possible to make use of the form as a guide to what points to cover, and then speak your piece on paper in a natural way.

Here are three formulas for letters. The first may be called the sales formula, the second the logical formula, and the third the rhetorical formula.

1. Get attention, provoke interest, rouse desire, obtain decision. Attention is curiosity fixed on something; interest is understanding of the nature and extent of what is new and its relationship to what is old; desire is the wish to take advantage of the proffered benefits; decision is based on confidence in what the writer says about his goods.

2. This is summarized: general, specific, conclusion. You start with a statement so broad and authoritative that it will not be disputed; you show that the general idea includes a specific idea; the conclusion is that what has been said about the general idea is also true of the specific.

3. This is very simple: picture, promise, prove, push. You write an attractive description of what you are selling; you promise that it will serve the reader well in such-and-such a way; you give examples of the commodity in use, proving that it has utility and worth; you urge the reader to take advantage of the promised values.

Selling needs ideas

Selling is done with ideas, so never throw away an idea even if it is of no use at the moment. Put it into your idea file where it will rub against other ideas and perhaps produce something new. The file is like an incubator. Thoughts and fancies you put into it will hatch out projects and plans.

Imagination helps in this operation. A correspondent of ordinary ability may never write anything that is not absolutely accurate and yet fail to interest his readers. This is a real weakness: to be perfect as to form but lacking in imagination and ideas.

Imagination should be given priority over judgment in preparing your first draft of a letter designed to sell. Then put reason to work: delete what is unnecessary, marshal your sentences into logical form so that your ideas advance in an orderly way; revise your words so that your thought is conveyed exactly as you wish it to be.

When you tell the advantages of your product or service or idea, and show how it will fill a need in the reader's life or job, in clear, truthful words placed in easily understood sentences brightened by ideas and imagination, you have done a good job of writing a sales letter.

Desire of the reader to do what you want done is created, just as in conversation, by both rational and emotional means, by proof and by persuasion, by giving reasons. Some goods and some buyers need nothing more than facts. An office manager buying pencils or pens for his staff will respond to an informative, factual, statistical sales letter. He is already sold on the idea of using pencils and pens, so you do not have to coach him about their usefulness; in fact, you may lose a sale if you give the impression of "teaching grandmother to suck eggs". What is needed is to catch his attention, give pertinent information about your product, and show him why buying from you will be profitable to him.

Try to make the information you give really enlightening. Comparing something unknown with something already known makes it possible to talk about the unknown. The analogy (like that between the heart and a pump) can be used as an aid in reasoning and in explaining or demonstrating.

The soft sell

The tone of a letter designed to sell something should be persuasive rather than insistent. It should seek to create a feeling of wanting, or at least an urge to "let's see".

People do not want to be told how to run their affairs, but anyone who shows them how to do things more economically or faster or better will find keen listeners. Soft sell gives the prospect credit for knowing a good thing when it is shown him, and acknowledges his right to make up his own mind.

The soft sell is recognition of the Missouri mule in human nature. Try to push a mule and he lashes out with his heels. Try to pull him by the halter rope and he braces his legs and defies you to budge him. "In the old cavalry," says A. C. Kemble in *Building Horsepower into Sales Letters*, "they said all it took to get a mule working for you was to recognize that he was an individualist who hated nagging and needed a chance to make up his own mind about things."

One hears a lot in advertising circles about "appeal". It is, according to the dictionary, "the power to attract, interest, amuse, or stimulate the mind or emotions."

Obviously, when you wish to influence someone you must take into account the kind of person you are addressing and what you want him to do. Your appeal must touch his feelings, needs and emotions. It strengthens your position if you can relate your own experience to that of the person you are addressing and write your message around the overlap of that experience.

The sort of mistake to be avoided with great care is slanting your appeal in a way that runs counter to the feelings of those whom you wish to influence.

It has been found in recent years that the advertising messages addressed to older people *as* older people did not win the desired response. In travel, for example, only a very small minority who want or need a sheltered situation are attracted by the semicustodial "trips for the elderly".

The Swiss Society for Market Research decided that "to sell anything to the over-65 age group it is important to keep one concept in mind: most senior citizens are vigorous and independent. Don't try to reach them with a head-on approach to the senior market. It probably won't work."

Writing a letter that pleases the recipient is not enough: it must be designed to lead to action. Do not fear to be explicit about what you want. Coyness in a letter is not attractive, and it exasperates the reader. Answer the reader's questions: "What has this to do with me?" and "Why should I do what this person is asking me to do?"

You may answer these questions and encourage a purchase by appealing to emotional motives like pride, innovation, emulation, or social prestige; or to rational motives like money gain, economy, security, time-saving or safety.

Read your letter critically

Imagine your letter to be your garden upon your return from vacation. You have to get into it and prune, clean up, tie up, and trim the edges.

Read the letter as if you were the recipient. How does it strike you? What can be added to attract attention? Is there anything irrelevant in it? Read the letter aloud to capture the conversational rhythm.

If you are not satisfied, do not crumple up the paper on which your draft is written. Try rearranging the paragraphs, the sentences, the words. Give the letter a new twist. Change the shape of your appeal. Delete anything that is distracting.

Be careful when trying to shorten a letter that seems to be too long. While a letter should be as short as possible, consistent with clearness and completeness, it is not the length that counts, but the depth. Since clearness and brevity sometimes get in the way of each other, remember that the right of way belongs to clearness. It will make a good impression if you find occasion to write: "I can be quite brief because this letter deals with a topic already well known to you."

The end of your letter, like the end of your pencil, should have a point. It should answer the reader's natural question: "So what?"

Follow through

Do not let your customer forget you. When you produce a piece of copy that hits the bull's-eye, that is not the time to sit back and take things easy. It is a time to imagine what you would do if you were in your competitor's chair . . . and then do it first.

Competition is a fact of life. Wherever there are two wild animals trying to live on the same piece of land or two persons depending upon the same source of sustenance, there is competition. The customer who was a prospect to you before he bought your goods is now a prospect to your competitor. With the proper follow-through attention he will turn to you when he needs up-dating of equipment or new goods.

Writing a letter that sells goods or ideas, and following through so as to retain the customer, requires just as much specialized talent and mental ability as any other kind of advertising, if not more.

When you run into difficulties, composition of the sales or follow-up letter may give you a feeling of confusion. You may feel like throwing up your hands in despair of finding the exactly right slant or the perfect array of words. That is not unnatural. Nietzsche, the German philosopher, said in *Thus Spake Zarathustra:* "I tell you, one must still have chaos in one, to give birth to a dancing star."

The effort is worthwhile. When you set yourself to snap out of the depressing pedestrian type of letter that is so commonplace, you are raising yourself and your firm to a place where people will sit up and pay attention. As a student of sales letter writing you will generate ideas, as a philosopher you will assess the letters as to their purpose and usefulness, and as a writer you will energize them.

To summarize: the backbone of the principles of writing letters that sell is made up of these vertebrae—know why you are writing and what about; believe in what you are writing; be tactful and friendly and truthful; base your appeal on the prospect's interests . . . and check your letter and revise it.

Letters can say no but keep or make friends

KERMIT ROLLAND

When he wrote this article, Kermit Rolland was a member of the Public Relations Department of the New York Life Insurance Company. He subsequently became a consultant specializing in better letter-writing programs.

An executive of a large New York business house has a unique collection of fan letters, each carefully mounted in a large leather album which he keeps close at hand as he runs through his mail each morning. Each letter in the growing collection is a testimonial to the man's talent for saying *No* and making a friend while doing so.

The letters were written to him in reply to *turn-down* letters he wrote for his company: refusals to extend credit, grant discounts, make adjustments, contribute to causes, or lengthen terms. Each of his original letters denied a request. Yet somehow each captured the good-will of a reader! Here are a few typical replies:

"My first reaction when I read your letter (refusing a discount) was to take my business elsewhere. But since thinking it over, I have decided to place my order anyway. You can credit for this the staff of psychologists you employ to write your letters for you."

"No one has ever said *No* to me before and made me like it."

"You made your reasons for refusal very clear and understandable. Thank you for having taken the trouble to do so. Another business house would just have said *No* and let it go at that."

"If more people would write as you did, there would be less suspicion on the part of the public toward the large corporations."

From *Printers' Ink,* 229 (October 7, 1949). Reprinted by permission of Frederic C. Decker for Decker Communications, Inc.

The man whose letters prompted this mail would hasten to explain that for every reader who will swallow a *No* and like it, there are two who will refuse to be moved by any appeal, no matter how effectively it may be presented. They will remain disappointed and disgruntled, though even one of these will occasionally accept a *No* albeit with bad grace.

"I think of my turn-down letters as a salvage operation," the writer explains. "I have found that an effective turn-down will often keep a friend who might otherwise have been lost through a stupid letter—and, surprisingly, a turn-down will sometimes create a new friendship."

Turn-down letters form a part of everyone's mail. In a large organization the bulk of the turn-down mail is handled by correspondents who often fail to realize the damaging chain reaction they may be setting off each time they send a flat *No* to a reader. The reader will rarely complain, it is true. But he will often chalk up an unfavorable impression of the company—and take his business elsewhere.

Long experience has shown one large company that the majority of its correspondents either lack the ability or are too pressed for time to take the trouble to compose an effective turn-down. It meets the problem by sending such letters on to its public relations department. The company feels that answering such correspondence is too important to be entrusted to anyone who has not had special experience in that field. For the bulk of the turn-downs, the PR people write guide letters to be used as patterns by the regular correspondents. But PR continues to handle all turn-downs that do not fit into the several categories of the guide letters. That its program is effective is attested by the fact that customers and others often write notes of appreciation for the thoughtful letters they have received.

Possibly no large corporation handled as many turn-downs during recent years as did the New York Telephone Company. Because of a shortage of equipment the company was unable to fill more than a fraction of its requests for new installations. People had to be put off, and they're still being put off. But so adroit are the turn-down letters the company sends out that most of the disappointed subscribers entertain no ill-will toward the company. Instead, they blame the war* for their disappointment, or material shortages, or a perverse fate—but rarely the company.

The secret of the telephone letters is that they *explain* to the reader why he can't have a telephone, and they explain so convincingly that the company retains the priceless good-will of the reader, even as he trudges to the neighbors to make a phone call.

In contrast to the successful technique used by the telephone people is the thoughtless approach of a utilities company whose turn-downs are so abrupt and so inconsiderate that more than one customer has thought wishfully of the day when bankruptcy could be visited upon it for its sins.

* *Editor's note:* World War II.

It is a maxim of modern correspondence practice that every letter should be a sales letter. It is relatively simple to accomplish this objective in routine letters of a positive nature. A turn-down letter is another matter. To be effective it does not seek to sell: it tries to buy. The commodity it bids for is the good-will of the reader.

The technique of an effective *No* letter is built upon an appeal to the reader. An effective turn-down is slanted to the reader's point of view; it emphasizes the fact that the reader is very important to the company. Here are a half-dozen tested methods for creating successful *No* letters:

1. Explain to your reader, if you can, the reason for the rejection

Nothing antagonizes a reader more than to be told that the granting of his request is not in line with company practice. This is a favorite evasion of many writers. It's easier to say "I regret to tell you that it is not the policy of the Company to . . ." or "The Company, has never . . . ." The company, not the writer, is to blame for the turn-down; or so many writers imply.

But these writers fail to realize that to the reader the writer is the company. So passing the buck will not work. It puts the writer and the company in the same hole as far as the reader is concerned. The reader's knowledge of a company is very often limited to the correspondence he exchanges with the individual or individuals who write the company's letters. The personality of the writer as it is reflected in a letter is the personality of the company on whose letterhead he is writing. Is it a friendly letter? Then it is a friendly company. Is it a cold letter? Then it is a cold company. Is it a stuffy letter? Then it is a plug-hat company.

Most readers are satisfied with a convincing explanation for a rejection. If the writer will ask himself, "Why can't we grant this request?" and then set down his answer is simple, direct, prose he will have solved half of his onerous problem.

Occasionally, it is true, it is difficult or awkward to tell the reader the precise reason for a refusal. If the reason is confidential, it requires a master touch to say *No* convincingly enough to leave the reader unruffled. A successful method in such situations is to assure the reader, and then reassure him, that his request has been thoroughly considered—and then screen the reason for the turn-down behind a generalization.

Here is a generalization which has proved effective in some instances: "We wish it were possible for us to grant every request for —————. Your request, I wish to assure you, was very carefully considered. But you will understand, I know, that it is not possible for us to say *Yes* in every instance. I regret that I must tell you that . . . ."

Another device which has been used with considerable success in handling a turn-down where it is not possible to give the reason is to hold the turn-down reply for a few days before mailing it. *It should be mentioned here that promptness is ordinarily a virtue in answering letters.* The delayed return letter opens with an explanation, not for the turn-down, *but for the delay in replying,* the implication being that the request was debated in the company for a number of days. Some magazine editors have used this device in rejecting manuscripts without being noticeably conscience-stricken.

2. Be sparing with apologies

An overly apologetic letter bears the stamp of insincerity. Few readers are so naïve as to believe that a company is really very upset about rejecting a request. It is very well to be "disturbed" or "distressed" when a *complaint* letter is being answered. But for an ordinary turn-down "I am sorry to tell you" or "I regret that we cannot" is usually sufficient. An explanation is more acceptable to a reader than profuse apologies. The reader is no dope. He knows the sound of wind when he hears it.

3. Try the *you* approach

The one person in the world who is important to the reader is himself. His interest in anything is directly linked to self-interest. He is really not very interested in the problems of the company, except so far as the problems affect him personally. This being the nature of the reader, it is wise to direct an appeal to him on his own terms. Convince the reader you have his interest at heart, and he will give you his good-will.

A reader who will stiffen at the sentence, "We give every request fair consideration," will relax when he meets the sentence rewritten, "*You* may be sure that we have given *your* request every consideration." Talk about the reader, not about the company.

4. But don't be afraid of *I*

While it is important to emphasize the *you* in a letter, it is not desirable to rule out every *we.* The company, although subordinate, is important, too. But let's examine this *we* for a moment.

For many years writers have been lurking behind the *we* of the company. Occasionally a very bold one would venture to call himself *the writer* or *the below signed writer.* It is difficult to understand this excessive timidity. If a letter is written by anyone with any kind of authority at all, why should he not make use of *I* instead of cloaking himself in the phony anonymity of *we*?

The use of the first person singular can be very effective in a turn-down letter. It establishes the important man-to-man atmosphere. "Take my word for it," the writer says in effect. "I am looking after your interests here."

I wish to assure you that we . . . I know you will understand that . . . I hope you will . . . I want you to know . . . I regret that I must tell you that we . . . You will appreciate, I know . . . Let me say that . . . etc., are phrases that have proved their effectiveness time and again.

5. Accentuate the positive

Occasionally it is possible to offer an alternative when denying a specific request. In this situation it is effective to offer the alternative goods or service at the beginning of the letter. Give the reader something positive at once if it is at all possible. Let the turn-down wait until the end of the letter when he has been cushioned for the shock.

6. If you can say *Yes,* say it at once. If you must say *No,* take a little longer

This rule-of-thumb is used successfully by one of the large philanthropic foundations. For each grant the foundation makes, it turns down a hundred appeals. So the bulk of the mail is made up of *No* letters.

The reason for the effectiveness of the "take a little longer" approach is explained by the underlying psychology of the third suggestion in this list: *the reader wants to feel important.* When he receives a *long* letter from an organization as well-known as the foundation, it puffs him up. It is apparent to him that some very busy and highly paid executive was sufficiently impressed by him and by his request to take the time to answer in full. And even though his request was denied, the stinger has been pulled.

A truly effective correspondence technique cannot, of course, be reduced to a half-dozen suggestions, and these suggestions are not offered as the basis for building a technique from scratch. But they have proved to be a useful supplement to an established correspondence method. They have salvaged good-will and good business that might otherwise have been lost.

How to write better memos

HAROLD K. MINTZ

Harold K. Mintz was Senior Technical Editor for the RCA Corporation when he wrote this article.

Memos—interoffice, intershop, interdepartmental—are the most important medium of in-house communication. This article suggests ways to help you sharpen your memos so that they will more effectively inform, instruct, and sometimes persuade your coworkers.

Memos are informal, versatile, free-wheeling. In-house they go up, down or sideways.* They can even go to customers, suppliers, and other interested outsiders. They can run to ten pages or more, but are mostly one to three pages. (Short memos are preferable. Typed single-space and with double-space between paragraphs, Lincoln's Gettysburg Address easily fits on one page, and the Declaration of Independence on two pages.) They can be issued on a one-shot basis or in a series, on a schedule or anytime at all. They can cover major or minor subjects.

Primary functions of memos encompass, but are not limited to:

- Informing people of a problem or situation.
- Nailing down responsibility for action, and a deadline for it.
- Establishing a file record of decisions, agreements and policies.

Secondary functions include:

- Serving as a basis for formal reports.
- Helping to bring new personnel up-to-date.
- Replacing personal contact with people you cannot get along with.
For example, the Shubert brothers, tyrannical titans of the American theatre for 40 years, often refused to talk to each other. They communicated by memo.

* We will return to this sentence later.

• Handling people who ignore your oral directions. Concerning the State Dept., historian Arthur Schlesinger quoted JFK as follows: "I have discovered finally that the best way to deal with State is to send over memos. They can forget phone conversations, but a memorandum is something which, by their system, has to be answered."

Memos can be used to squelch unjustified time-consuming requests. When someone makes what you consider to be an unwarranted demand or request, tell him to put it in a memo—just for the record. This tactic can save you much time.

Organization of the memo

Memos and letters are almost identical twins. They differ in the following ways: Memos normally remain in-house, memos don't usually need to "hook" the reader's interest, and memos covering a current situation can skip a background treatment.

Overall organization of a memo should ensure that it answer three basic questions concerning its subject:

1. What are the facts?
2. What do they mean?
3. What do we do now?

To supply the answers, a memo needs some or all of the following elements: summary, conclusions and recommendations, introduction, statement of problem, proposed solution, and discussion. Incidentally, these elements make excellent headings to break up the text and guide the readers.

In my opinion, every memo longer than a page should open with a summary, preferably a short paragraph. Thus, recipients can decide in seconds whether they want to read the entire memo.

Two reasons dictate placing the summary at the very beginning. There, of all places, you have the reader's undivided attention. Second, readers want to know, quickly, the meaning or significance of the memo.

Obviously, a summary cannot provide all the facts (Question 1, above) but it should capsule their meaning, and highlight a course of action.

When conclusions and recommendations are not applicable, forget them. When they are, however, you can insert them either right after the abstract or at the end of the memo. Here's one way to decide: If you expect readers to be neutral or favorable toward your conclusions and recommendations, put them up front. If you expect a negative reaction, put them at the end. Then, conceivably, your statement of the problem and your discussion of it may swing readers around to your side by the time they reach the end.

The introduction should give just enough information for the readers to be able to understand the statement of the problem and its discussion.

Literary qualities

A good memo need not be a Pulitzer Prize winner, but it does need to be clear, brief, relevant. LBJ got along poorly with his science adviser, Donald Hornig, because Hornig's memos, according to a White House staffer, "were terribly long and complicated. The President couldn't read through a page or two and understand what Don wanted him to do, so he'd send it out to us and ask us what it was all about. Then we'd put a short cover-memo on top of it and send it back in. The President got mad as hell at long memos that didn't make any sense."

Clarity is paramount. Returning to the asterisked sentence in the second paragraph of the introduction, I could have said: "Memoranda are endowed with the capability of internal perpendicular and lateral deployment." Sheer unadulterated claptrap.

To sum up, be understandable and brief, but not brusque, and get to the point.

Another vitally important trait is a personal, human approach. Remember that your memos reach members of your own organization; that's a common bond worth exploiting. Your memos should provide them with the pertinent information they need (no more and no less) and in the language they understand. Feel free to use people's names, and personal pronouns and adjectives: you-your, we-our, I-mine. Get people into the act; it's they who do the work.

Lastly, a well-written memo should reflect diplomacy or political savvy. More than once, Hornig's memos lighted the fuse of LBJ's temper. One memo, regarded as criticizing James E. Webb (then the head of NASA), LBJ's friend, infuriated the President.

Another example of a politically naive memo made headlines in England three years ago. A hospital superintendent wrote a memo to his staff, recommending that aged and chronically ill patients should not be resuscitated after heart failure. Public reaction exploded so overwhelmingly against the superintendent that shock waves even shook Prime Minister Wilson's cabinet. Result? The Health Ministry torpedoed the recommendation.

Two other courses of action would have been more tactful for the superintendent: make the recommendation orally to his staff or, if he insisted on a memo, stamp it "private" and distribute it accordingly.

Literary style is a nebulous subject, difficult to pin down. Yet if you develop a clear, taut way of writing, you may end up in the same happy predicament as Lawrence of Arabia. He wrote "a violent memorandum" on a British-Arab problem, a memo whose "acidity and force" so impressed the commanding general that he wired it to London. Lawrence noted in his "Seven Pillars of Wisdom" that, "My popularity with the military staff in Egypt, due to the sudden help I had lent . . . was novel and rather amusing. They began to be polite to me, and to say that I was observant, with a pungent style. . . ."

Format of the memo

Except for minor variations, the format to be used is standard. The memo dispenses with the addresses, salutations, and complimentary closes used in letters. Although format is a minor matter, it does rate some remarks.

To and From Lines—Names and departments are enough.

Subject—Capture its essence in ten words or less. Any subject that drones on for three or four lines may confuse or irritate readers.

Distribution—Send the memo only to people involved or interested in the subject matter. If they number less than say, ten, list them alphabetically on page 1; if more than ten, put them at the end.

Text—Use applicable headings listed after the three questions under "Organization."

Paragraphs—If numbering or lettering them helps in any way, do it.

Line Spacing—Single space within paragraphs, and double space between.

Underlines and Capitals—Used sparingly, they emphasize important points.

Number of Pages—Some companies impose a one-page limit, but it's an impractical restriction because some subjects just won't fit on one page. As a result, the half-baked memo requires a second or third memo to beef it up.

Figures and Tables—Use them; they'll enhance the impact of your memos.

Conclusion

Two cautions are appropriate. First, avoid writing memos that baffle people, like the one that Henry Luce once sent to an editor of *Time*. "There are only 30,000,000 sheep in the U.S.A.—same as 100 years ago. What does this prove? Answer???"

Second, avoid "memo-itis," the tendency to dash off memos at the drop of a pen, especially to the boss. In his book, "With Kennedy," Pierre Salinger observed that "a constant stream of memoranda" from Professor Arthur Schlesinger caused JFK to be "impatient with their length and frequency."

Clear writing means clear thinking means . . .

MARVIN H. SWIFT

When he wrote this article, Marvin H. Swift was Associate Professor of Communication at the General Motors Institute.

*I*f you are a manager, you constantly face the problem of putting words on paper. If you are like most managers, this is not the sort of problem you enjoy. It is hard to do, and time consuming; and the task is doubly difficult when, as is usually the case, your words must be designed to change the behavior of others in the organization.

But the chore is there and must be done. How? Let's take a specific case.

Let's suppose that everyone at X Corporation, from the janitor on up to the chairman of the board, is using the office copiers for personal matters; income tax forms, church programs, children's term papers, and God knows what else are being duplicated by the gross. This minor piracy costs the company a pretty penny, both directly and in employee time, and the general manager—let's call him Sam Edwards—decides the time has come to lower the boom.

Sam lets fly by dictating the following memo to his secretary:

```
To: All Employees
From: Samuel Edwards, General Manager
Subject: Abuse of Copiers

It has recently been brought to my attention that many of the
people who are employed by this company have taken advantage
of their positions by availing themselves of the copiers. More
specifically, these machines are being used for other than
company business.
```

Obviously, such practice is contrary to company policy and
must cease and desist immediately. I wish therefore to inform
all concerned--those who have abused policy or will be abusing
it--that their behavior cannot and will not be tolerated.
Accordingly, anyone in the future who is unable to control
himself will have his employment terminated.

If there are any questions about company policy, please feel
free to contact this office.

Now the memo is on his desk for his signature. He looks it over; and the more
he looks, the worse it reads. In fact, it's lousy. So he revises it three times, until
it finally is in the form that follows:

To: All Employees
From: Samuel Edwards, General Manager
Subject: Use of Copiers

We are revamping our policy on the use of copiers for personal
matters. In the past we have not encouraged personnel to use
them for such purposes because of the costs involved. But we
also recognize, perhaps belatedly, that we can solve the
problem if each of us pays for what he takes.

We are therefore putting these copiers on a pay-as-you-go
basis. The details are simple enough . . . .

Samuel Edwards

This time Sam thinks the memo looks good, and it *is* good. Not only is the
writing much improved, but the problem should now be solved. He therefore
signs the memo, turns it over to his secretary for distribution, and goes back to
other things.

From verbiage to intent

I can only speculate on what occurs in a writer's mind as he moves from a poor
draft to a good revision, but it is clear that Sam went through several specific
steps, mentally as well as physically, before he had created his end product:

- He eliminated wordiness.
- He modulated the tone of the memo.
- He revised the policy it stated.

Let's retrace his thinking through each of these processes.

Eliminating wordiness

Sam's basic message is that employees are not to use the copiers for their own affairs at company expense. As he looks over his first draft, however, it seems so long that this simple message has become diffused. With the idea of trimming the memo down, he takes another look at his first paragraph:

It has recently been brought to my attention that many of the people who are employed by this company have taken advantage of their positions by availing themselves of the copiers. More specifically, these machines are being used for other than company business.

He edits it like this:

ITEM: "recently"
COMMENT TO HIMSELF: Of course; else why write about the problem? So delete the word.

ITEM: "It has been brought to my attention"
COMMENT: Naturally. Delete it.

ITEM: "the people who are employed by this company"
COMMENT: Assumed. Why not just "employees"?

ITEM: "by availing themselves" and "for other than company business"
COMMENT: Since the second sentence repeats the first, why not coalesce?

And he comes up with this:

Employees have been using the copiers for personal matters.

He proceeds to the second paragraph. More confident of himself, he moves in broader swoops, so that the deletion process looks like this:

Obviously, such practice is contrary to company policy and ~~must cease and desist immediately. i wish therefore to inform all concerned — those who have abused policy or will be abusing it that their behavior cannot and will not be tolerated. Accordingly anyone in the future who is unable to control himself will have his employment terminated.~~ will result in dismissal.

The final paragraph, apart from "company policy" and "feel free," looks all right, so the total memo now reads as follows:

```
To: All Employees
From: Samuel Edwards, General Manager
Subject: Abuse of Copiers
```

```
Employees have been using the copiers for personal matters.
Obviously, such practice is contrary to company policy and will
result in dismissal.
```

```
If there are any questions, please contact this office.
```

Sam now examines his efforts by putting these questions to himself:

QUESTION: Is the memo free of deadwood?
ANSWER: Very much so. In fact, it's good, tight prose.
QUESTION: Is the policy stated?
ANSWER: Yes—sharp and clear.
QUESTION: Will the memo achieve its intended purpose?
ANSWER: Yes. But it sounds foolish.
QUESTION: Why?
ANSWER: The wording is too harsh; I'm not going to fire anybody over this.
QUESTION: How should I tone the thing down?

To answer this last question, Sam takes another look at the memo.

Correcting the tone

What strikes his eye as he looks it over? Perhaps these three words:

- Abuse . . .
- Obviously . . .
- . . . dismissal . . .

The first one is easy enough to correct: he substitutes "use" for "abuse." But "obviously" poses a problem and calls for reflection. If the policy is obvious, why are the copiers being used? Is it that people are outrightly dishonest? Probably not. But that implies the policy isn't obvious; and whose fault is this? Who neglected to clarify policy? And why "dismissal" for something never publicized?

These questions impel him to revise the memo once again:

```
To: All Employees
From: Samuel Edwards, General Manager
Subject: Use of Copiers
```

```
Copiers are not to be used for personal matters. If there are any
questions, please contact this office.
```

Revising the policy itself

The memo now seems courteous enough—at least it is not discourteous—but it is just a blank, perhaps overly simple, statement of policy. Has he really thought through the policy itself?

Reflecting on this, Sam realizes that some people will continue to use the copiers for personal business anyhow. If he seriously intends to enforce the basic policy (first sentence), he will have to police the equipment, and that raises the question of costs all over again.

Also, the memo states that he will maintain an open-door policy (second sentence)—and surely there will be some, probably a good many, who will stroll in and offer to pay for what they use. His secretary has enough to do without keeping track of affairs of that kind.

Finally, the first and second sentences are at odds with each other. The first says that personal copying is out, and the second implies that it can be arranged.

The facts of organizational life thus force Sam to clarify in his own mind exactly what his position on the use of copiers is going to be. As he sees the problem now, what he really wants to do is put the copiers on a pay-as-you-go basis. After making that decision, he begins anew:

```
To: All Employees
From: Samuel Edwards, General Manager
Subject: Use of Copiers

We are revamping our policy on the use of copiers . . . . . .
```

This is the draft that goes into distribution and now allows him to turn his attention to other problems.

The chicken or the egg?

What are we to make of all this? It seems a rather lengthy and tedious report of what, after all, is a routine writing task created by a problem of minor importance. In making this kind of analysis, have I simply labored the obvious?

To answer this question, let's drop back to the original draft. If you read it over, you will see that Sam began with this kind of thinking:

- "The employees are taking advantage of the company."

- "I'm a nice guy, but now I'm going to play Dutch uncle."
∴ "I'll write them a memo that tells them to shape up or ship out."

In his final version, however, his thinking is quite different:

- "Actually, the employees are pretty mature, responsible people. They're capable of understanding a problem."
- "Company policy itself has never been crystallized. In fact, this is the first memo on the subject."
- "I don't want to overdo this thing—any employee can make an error in judgment."
- ∴ "I'll set a reasonable policy and write a memo that explains how it ought to operate."

Sam obviously gained a lot of ground between the first draft and the final version, and this implies two things. First, if a manager is to write effectively, he needs to isolate and define, as fully as possible, all the critical variables in the writing process and scrutinize what he writes for its clarity, simplicity, tone, and the rest. Second, after he has clarified his thoughts on paper, he may find that what he has written is not what has to be said. In this sense, writing is feedback and a way for the manager to discover himself. What are his real attitudes toward that amorphous, undifferentiated gray mass of employees "out there"? Writing is a way of finding out. By objectifying this thoughts in the medium of language, he gets a chance to see what is going on in his mind.

In other words, *if the manager writes well, he will think well.* Equally, the more clearly he has thought out his message before he starts to dictate, the more likely he is to get it right on paper the first time round. In other words, *if he thinks well, he will write well.*

Hence we have a chicken-and-the-egg situation: writing and thinking go hand in hand; and when one is good, the other is likely to be good.

Revision sharpens thinking

More particularly, rewriting is the key to improved thinking. It demands a real openmindedness and objectivity. It demands a willingness to cull verbiage so that ideas stand out clearly. And it demands a willingness to meet logical contradictions head on and trace them to the premises that have created them. In short, it forces a writer to get up his courage and expose his thinking process to his own intelligence.

Obviously, revising is hard work. It demands that you put yourself through the wringer, intellectually and emotionally, to squeeze out the best you can offer. Is it worth the effort? Yes, it is—if you believe you have a responsibility to think and communicate effectively.

4

Reports and proposals

Audience analysis: the problem and a solution

J. C. MATHES and
DWIGHT W. STEVENSON

Both J. C. Mathes and Dwight W. Stevenson are
professors of humanities in the College of
Engineering at the University of Michigan.

*E*very communication situation involves three fundamental components: a writer, a message, and an audience. However, many report writers treat the communication situation as if there were only two components: a writer and his message. Writers often ignore their readers because writers are preoccupied with their own problems and with the subject matter of the communication. The consequence is a poorly designed, ineffective report.

As an example, a student related to the class her first communication experience on a design project during summer employment with an automobile company. After she had been working on her assignment for a few weeks, her supervisor asked her to jot him a memo explaining what she was doing. Not wanting to take much time away from her work and not thinking the report very important, she gave him a handwritten memo and continued her technical activities. Soon after, the department manager inquired on the progress of the project. The supervisor immediately responded that he had just had a progress report, and thereupon forwarded the engineer's brief memo. Needless to say, the engineer felt embarrassed when her undeveloped and inadequately explained memo became an official report to the organization. The engineer thought her memo was written just to her supervisor, who was quite familiar with her assignment. Due to her lack of experience with organizational behavior, she made several false assumptions about her report audience, and therefore about her report's purpose.

The inexperienced report writer often fails to design his report effectively because he makes several false assumptions about the report writing situation.

If the writer would stop to analyze the audience component he would realize that:

1. It is false to assume that the person addressed is the audience.

2. It is false to assume that the audience is a group of specialists in the field.

3. It is false to assume that the report has a finite period of use.

4. It is false to assume that the author and the audience always will be available for reference.

5. It is false to assume that the audience is familiar with the assignment.

6. It is false to assume that the audience has been involved in daily discussions of the material.

7. It is false to assume that the audience awaits the report.

8. It is false to assume that the audience has time to read the report.

Assumptions one and two indicate a writer's lack of awareness of the nature of his report audience. Assumptions three, four, and five indicate his lack of appreciation of the dynamic nature of the system. Assumptions six, seven, and eight indicate a writer's lack of consideration of the demands of day-by-day job activity.

A report has value only to the extent that it is useful to the organization. It is often used primarily by someone other than the person who requested it. Furthermore, the report may be responding to a variety of needs within the organization. These needs suggest that the persons who will use the report are not specialists or perhaps not even technically knowledgeable about the report's subject. The specialist is the engineer. Unless he is engaged in basic research, he usually must communicate with persons representing many different areas of operation in the organization.

In addition, the report is often useful over an extended period of time. Each written communication is filed in several offices. Last year's report can be incomprehensible if the writer did not anticipate and explain his purpose adequately. In these situations, even within the office where a report originated, the author as well as his supervisor will probably not be available to explain the report. Although organizational charts remain unchanged for years, personnel, assignments, and professional roles change constantly. Because of this dynamic process, even the immediate audience of a report sometimes is not familiar with the writer's technical assignment. Thus, the report writer usually must design his report for a dynamic situation.

Finally, the report writer must also be alert to the communication traps in relatively static situations. Not all readers will have heard the coffee break chats that fill in the details necessary to make even a routine recommendation convincing. A report can arrive at a time when the reader's mind is churning with other concerns. Even if it is expected, the report usually meets a reader who needs to act immediately. The reader usually does not have time to read through the whole report; he wants the useful information clearly and succinctly. To the reader, time probably is the most important commodity. Beginning report writers seldom realize they must design their reports to be used efficiently rather than read closely.

The sources of the false assumptions we have been discussing are not difficult to identify. The original source is the artificial communication a student is required to perform in college. In writing only for professors, a student learns to write for audiences of one, audiences who know more than the writer knows, and audiences who have no instrumental interests in what the report contains. The subsequent source, on the job, is the writer's natural attempt to simplify his task. The report writer, relying upon daily contact and familiarity, simply finds it easier to write a report for his own supervisor than to write for a supervisor in a different department. The writer also finds it easier to concentrate upon his own concerns than to consider the needs of his readers. He finds it difficult to address complex audiences and face the design problems they pose.

Audience components and problems they pose

To write a report you must first understand how your audience poses a problem. Then you must analyze your audience in order to be able to design a report structure that provides an optimum solution. To explain the components of the report audience you must do more than just identify names, titles, and roles. You must determine who your audiences are as related to the purpose and content of your report. "Who" involves the specific operational functions of the persons who will read the report, as well as their educational and business backgrounds. These persons can be widely distributed, as is evident if you consider the operational relationships within a typical organization.

Classifying audiences only according to directions of communication flow along the paths delineated by the conventional organizational chart, we can identify three types of report audiences: *horizontal, vertical,* and *external.* For example, in the organization chart in Figure 1, *Part of organization chart for naval ship engineering center,*[1] horizontal audiences exist on each level. The Ship Concept Design Division and the Command and Surveillance Division form horizontal audiences for each other. Vertical audiences exist between levels. The Ship Concept Design Division and the Surface Ship Design Branch form vertical audiences for each other. External audiences exist when any unit interacts with a separate organization, such as when the Surface Ship Design Branch communicates with the Newport News Shipbuilding Company.

[1] A reference in H. B. Benford and J. C. Mathes, *Your Future in Naval Architecture,* Richards Rosen, New York, 1968.

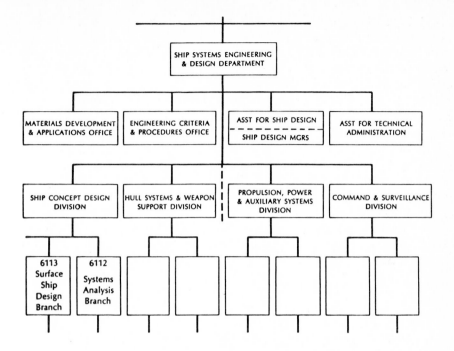

Fig. 1. Part of organization chart for naval ship engineering center.

What the report writer first must realize is the separation between him and any of these three types of audiences. Few reports are written for horizontal audiences within the same unit, such as from one person in the Surface Ship Design Branch to another person or project group within the Surface Ship Design Branch itself. Instead, a report at least addresses horizontal audiences within a larger framework, such as from the Surface Ship Design Branch to the Systems Analysis Branch. Important reports usually have complex audiences, that is, vertical and horizontal, and sometimes external audiences as well.

An analysis of the problems generated by horizontal audiences—often assumed to pose few problems—illustrates the difficulties most writers face in all report writing situations. A systems engineer in the Systems Analysis Branch has little technical education in common with the naval architect in the Surface Ship Design Branch. In most colleges he takes only a few of the same mathematics and engineering science courses. The systems engineer would not know the wave resistance theory familiar to the naval architect, although he could use the results of his analysis. In turn, the naval architect would not know stochastics and probability theory, although he could understand systems models. But the differences between these audiences and writers go well beyond differences in training. In addition to having different educational backgrounds, the audiences will have different concerns, such as budget, production, or contract obligations. The audiences will also be separated from the writer by organizational politics and competition, as well as by personality differences among the people concerned.

When the writer addresses a horizontal audience in another organizational unit, he usually addresses a person in an organizational role. When addressed to the role rather than the person, the report is aimed at a department or a group. This means the report will have audiences in addition to the person addressed. It may be read primarily by staff personnel and subordinates. The addressee ultimately may act on the basis of the information reported, but at times he serves only to transfer the report to persons in his department who will use it. Furthermore, the report may have audiences in addition to those in the department addressed. It may be forwarded to other persons elsewhere, such as lawyers and comptrollers. The report travels routinely throughout organizational paths, and will have unknown or unanticipated audiences as well.

Consequently, even when on the same horizontal organizational level, the writer and his audience have little in common beyond the fact of working for the same organization, of having the same "rank," and perhaps of having the same educational level of attainment. Educational backgrounds can be entirely different; more important, needs, values, and uses are different. The report writer may recommend the choice of one switch over another on the basis of a cost-efficiency analysis; his audiences may be concerned for business relationships, distribution patterns, client preferences, and budgets. Therefore, the writer should not assume that his audience has technical competence in the field, familiarity with the technical assignment, knowledge of him or of personnel in his group, similar value perspectives, or even complementary motives. The differences between writer and audience are distinctive, and may even be irreconcilable.

The differences are magnified when the writer addresses vertical audiences. Reports directed at vertical audiences, that is, between levels of an organization chart, invariably have horizontal audience components also. These complex report writing situations pose significant communication problems for the writer. Differences between writer and audience are fundamental. The primary audiences for the reports, especially informal reports, must act or make decisions on the basis of the reports. The reports thus have only instrumental value, that is, value insofar as they can be used effectively. The writer must design his report primarily according to how it will be used.

In addition to horizontal audiences and to vertical audiences, many reports are also directed to external audiences. External audiences, whether they consist of a few or many persons, have the distinctive, dissimilar features of the complex vertical audience. With external audiences these features invariably are exaggerated, especially those involving need and value. An additional complication is that the external audience can judge an entire organization on the basis of the writer's report. And sometimes most important of all, concerns for tact and business relationships override technical concerns.

In actual practice the writer often finds audiences in different divisions of his own company to be "external" audiences. One engineer encountered this problem in his first position after graduation. He was sent to investigate the inconsistent test data being sent to his group from a different division of the company in another city. He found that the test procedures being used in that division were faulty. However, at his supervisor's direction he had to write a report that would not "step on any toes." He had to write the report in such a manner as to have the other division correct its test procedures while not implying that the division was in any way at fault. An engineer who assumes that the purpose of his report is just to explain a technical investigation is poorly prepared for professional practice.

Most of the important communication situations for an engineer during his first five years out of college occur when he reports to his supervisor, department head, and beyond. In these situations, his audiences are action-oriented line management who are uninterested in the technical details and may even be unfamiliar with the assignment. In addition, his audiences become acquainted with him professionally through his reports; therefore, it is more directly the report than the investigation that is important to the writer's career.

Audience components and the significant design problems they pose are well illustrated by the various audiences for a formal report written by an engineer on the development of a process to make a high purity chemical, as listed in Figure 2, *Complex audience components for a formal report by a chemical engineer on a process to make a high purity chemical.* The purpose of

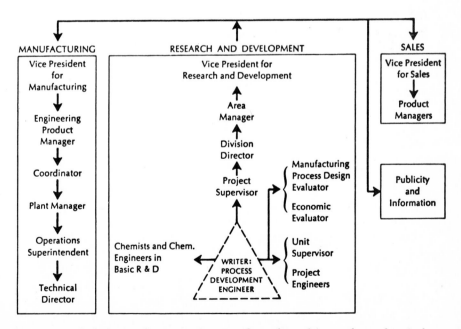

Fig. 2. Complex audience components for a formal report by a chemical engineer on a process to make a high purity chemical.

the report was to explain the process; others would make a feasibility study of the process and evaluate it in comparison to other processes.

The various audiences for this report, as you can determine just by reading their titles, would have had quite different roles, backgrounds, interests, values, needs, and uses for the report. The writer's brief analysis of the audiences yielded the following:

> He could not determine the nature of many of his audiences, who they were, or what the specifics of their roles were.
>
> His audiences had little familiarity with his assignment.
>
> His report would be used for information, for evaluation of the process, and for evaluation of the company's position in the field.
>
> Some of his audiences would have from a minute to a half hour to glance at the report, some would take the report home to study it, and some would use it over extended periods of time for process analysis and for economic and manufacturing feasibility studies.
>
> The useful lifetime of the report could be as long as twenty years.
>
> The report would be used to evaluate the achievements of the writer's department.
>
> The report would be used to evaluate the writing and technical proficiencies of the writer himself.

This report writer classified his audiences in terms of the conventional organization chart. Then to make them more than just names, titles, and roles he asked himself what they would know about his report and how they would use it. Even then he had only partially solved his audience problem and had just begun to clarify the design problems he faced. To do so he needed to analyze his audiences systematically.

A method for systematic audience analysis

To introduce the audience problem that report writers must face, we have used the conventional concept of the organization chart to classify audiences as *horizontal, vertical,* and *external.* However, when the writer comes to the task of performing an instrumentally useful audience analysis for a particular report, this concept of the organization and this classification system for report audiences are not very helpful.

First, the writer does not view from outside the total communication system modeled by the company organization chart. He is within the system himself, so his view is always relative. Second, the conventional outsider's view does not yield sufficiently detailed information about the report audiences. A single bloc on the organization chart looks just like any other bloc, but in fact each bloc represents one or several human beings with distinctive roles, backgrounds, and personal characteristics. Third, and most importantly, the outsider's view does not help much to clarify the specific routes of communication, as determined by audience needs, which an individual report will follow. The organization chart may describe the organization, but it does not describe how the organization functions. Thus many of the routes a report follows—and consequently the needs it addresses—will not be signaled by the company organization chart.

In short, the conventional concept of report audiences derived from organization charts is necessarily abstract and unspecific. For that reason a more effective method for audience analysis is needed. In the remaining portion of this chapter we will present a three-step procedure. The procedure calls for preparing an egocentric organization chart to identify individual report readers, characterizing these readers, and classifying them to establish priorities. Based upon an egocentric view of the organization and concerned primarily with what report readers need, this system should yield the information the writer must have if he is to design an individual report effectively.

Prepare an egocentric organization chart

An egocentric organization chart differs from the conventional chart in two senses. First, it identifies specific individuals rather than complex organizational units. A bloc on the conventional chart may often represent a number of people, but insofar as possible the egocentric chart identifies particular individuals who are potential readers of reports a writer produces. Second, the egocentric chart categorizes people in terms of their proximity to the report writer rather than in terms of their hierarchical relationship to the report writer. Readers are not identified as organizationally superior, inferior, or equal to the writer, but rather as near or distant from the writer. We find it effective to identify four different degrees of distance as is illustrated in Figure 3, *Egocentric organization chart*. In this figure, with the triangle representing the writer, each circle is an individual reader identified by his organizational title and by his primary operational concerns. The four degrees of distance are identified by the four concentric rings. The potential readers in the first ring are those people with whom the writer associates daily. They are typically those people in his same office or project group. The readers in the second ring are those people in other offices with whom the writer must normally interact in order to perform his job. Typically, these are persons in adjacent and management groups. The readers in the third ring are persons relatively more distant but still within the same organization. They are distant management, public relations, sales, legal department, production, purchasing, and so on. They are operationally dissimilar persons. The readers in the fourth ring are persons beyond the organization. They may work for the same company but in a division in another city. Or they may work for an entirely different organization.

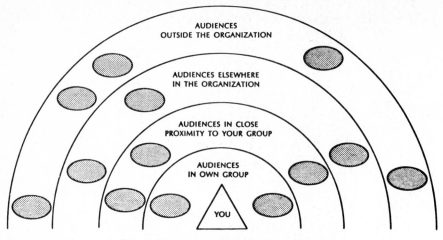

Fig. 3. Egocentric organization chart.

Having prepared the egocentric organization chart, the report writer is able to see himself and his potential audiences from a useful perspective. Rather than seeing himself as an insignificantly small part of a complex structure—as he is apt to do with the conventional organizational chart—the writer sees himself as a center from which communication radiates throughout an organization. He sees his readers as individuals rather than as faceless blocs. And he sees that what he writes is addressed to people with varying and significant degrees of difference.

A good illustration of the perspective provided by the egocentric organization chart is the chart prepared by a chemical engineer working for a large corporation, Figure 4, *Actual egocentric organization chart of an engineer in a large corporation.* It is important to notice how the operational concerns of the persons even in close proximity vary considerably from those of the development engineer. What these people need from reports written by this engineer, then, has little to do with the processes by which he defined his technical problems.

The chemical engineer himself is concerned with the research and development of production processes and has little interest in, or knowledge of, budgetary matters. Some of the audiences in his group are chemists concerned with production—not with research and development. Because of this they have, as he said, "lost familiarity with the technical background, and instead depend mostly on experience." Other audiences in his group are technicians concerned only with operations. With only two years of college, they have had no more than introductory chemistry courses and have had no engineering courses.

Still another audience in his group is his group leader. Rather than being concerned with development, this reader is concerned with facilities and production operations. Consequently, he too is "losing familiarity with the technical material." Particularly significant for the report writer is that his group

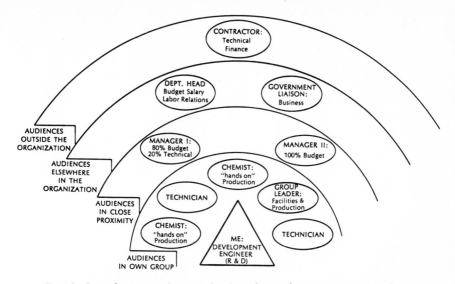

Fig. 4. Actual egocentric organization chart of an engineer in a large corporation.

leader in his professional capacity does not use his B.S.Ch.E. degree. His role is that of manager, so his needs have become administrative rather than technical.

The concerns of the chemical engineer/report writer's audiences in close proximity to his group change again. Instead of being concerned with development or production operations, these audiences are primarily concerned with the budget. They have little technical contact, and are described as "business oriented." Both Manager I and Manager II are older, and neither has a degree in engineering. One has a Ph.D. degree in chemistry, the other an M.S. degree in technology. Both have had technical experience in the lab, but neither can readily follow technical explanations. As the chemical engineer said, both would find it "difficult to return to the lab."

The report writer's department head and the other persons through whom the group communicates with audiences elsewhere in the organization, and beyond it, have additional concerns as well as different backgrounds. The department head is concerned with budget, personnel, and labor relations. The person in contact with outside funding units—in this case, a government agency—has business administration degrees and is entirely business oriented. The person in contact with subcontractors has both technical and financial concerns.

Notice that when this writer examined his audiences even in his own group as well as those in close proximity to him, he saw that the natures, the backgrounds, and especially the operational concerns of his audiences vary and differ considerably. As he widened the scope of his egocentric organization chart, he knew less and less about his audiences. However, he could assume they will vary even more than those of the audiences in close proximity.

Thus, in the process of examining the audience situation with an egocentric organization chart, a report writer can uncover not only the fact that audiences have functionally different interests, but also the nature of those functional differences. He can proceed to classify the audiences for each particular report in terms of audience needs.

Preparation of the egocentric organization chart is the first step of your procedure of systematic audience analysis. Notice that this step can be performed once to describe your typical report audience situation but must be particularized for each report to define the audiences for that report. Having prepared the egocentric chart once, the writer revises his chart for subsequent reports by adding or subtracting individual audiences.

Characterize the individual report readers

In the process of preparing the egocentric organization chart, you immediately begin to think of your individual report readers in particular terms. In preparing the egocentric chart discussed above, the report writer mentioned such items as a reader's age, academic degrees, and background in the organization as well as his operational concerns. All of these particulars will come to mind when you think of your audiences as individuals. However, a systematic rather than piecemeal audience analysis will yield more useful information. The second step of audience analysis is, therefore, a systematic characterization of each person identified in the egocentric organization chart. A systematic characterization is made in terms of *operational, objective,* and *personal* characteristics.

The *operational characteristics* of your audiences are particularly important. As you identify the operational characteristics for a person affected by your report, try to identify significant differences between his or her role and yours. What are his professional values? How does he spend his time? That is, will his daily concerns and attitudes enable him to react to your report easily, or will they make it difficult for him to grasp what you are talking about? What does he know about your role, and in particular, what does he know or remember about your technical assignment and the organizational problem that occasioned your report to come to him? You should also consider carefully what he will need from your report. As you think over your entire technical investigation, ask yourself if that person will involve staff personnel in action on your report, or if he will in turn activate other persons elsewhere in the organization when he receives the report. If he should, you must take their reactions into account when you write your report.

In addition, you should ask yourself, "How will my report affect his role?" A student engineer recently told us of an experience he had during summer employment when he was asked to evaluate the efficiency of the plant's waste treatment process. Armed with his fresh knowledge from advanced chemical engineering courses, to his surprise he found that, by making a simple change in the process, the company could save more than $200,000 a year. He fired off his report with great anticipation of glowing accolades—none came. How had his report affected the roles of some of his audiences? Although the writer had not considered the report's consequences when he wrote it, the supervisor, the manager, and related personnel now were faced with the problem of accounting for their waste of $200,000 a year. It should have been no surprise that they were less than elated over his discovery.

By *objective characteristics* we mean specific, relevant background data about the person. As you try to identify his or her educational background, you may note differences you might have otherwise neglected. Should his education seem to approximate yours, do not assume he knows what you know. Remember that the half-life of engineering education today is about five years. Thus, anyone five to ten years older than you, if you are recently out of college, probably will be only superficially familiar with the material and jargon of your advanced technical courses. If you can further identify his past professional experiences and roles, you might be able to anticipate his first-hand knowledge of your role and technical activities as well as to clarify any residual organizational commitments and value systems he might have. When you judge his knowledge of your technical area, ask yourself, "Could he participate in a professional conference in my field of specialization?"

For *personal characteristics*, when you identify a person by name, ask yourself how often the name changes in this organizational role. When you note his or her approximate age, remind yourself how differences in age can inhibit communication. Also note personal concerns that could influence his reactions to your report.

A convenient way to conduct the audience analysis we have been describing and to store the information it yields is to use an analysis form similar to the one in Figure 5, *Form for characterizing individual report readers.* It may be a little time-consuming to do this the first time around, but you can establish a file of audience characterizations. Then you can add to or subtract from this file as an individual communication situation requires.

One final point: This form is a means to an end rather than an end in itself. What is important for the report writer is that he think systematically about the questions this form raises. The novice usually has to force himself to analyze his audiences systematically. The experienced writer does this automatically.

Classify audiences in terms of how they will use your report

For each report you write, trace out the communication routes on your egocentric organization chart and add other routes not on the chart. Do not limit these routes to those specifically identified by the assignment and the addressees of the report. Rather, think through the total impacts of your report on the or-

NAME: TITLE:

A. OPERATIONAL CHARACTERISTICS:
   1. His role within the organization and consequent value system:

   2. His daily concerns and attitudes:

   3. His knowledge of your technical responsibilities and assignment:

   4. What he will need from your report:

   5. What staff and other persons will be activitated by your report through him:

   6. How your report could affect his role:

B. OBJECTIVE CHARACTERISTICS:
   1. His education—levels, fields, and years:

   2. His past professional experiences and roles:

   3. His knowledge of your technical area:

C. PERSONAL CHARACTERISTICS:
   Personal characteristics that could influence his reactions—-
   age, attitudes, pet concerns, etc.

Fig. 5. Form for characterizing individual report readers.

ganization. That is, think in terms of the first, second, and even some third-order consequences of your report, and trace out the significant communication routes involved. All of these consequences define your actual communication.

When you think in terms of consequences, primarily you think in terms of the uses to which your report will be put. No longer are you concerned with your technical investigation itself. In fact, when you consider how readers will use your report, you realize that very few of your potential readers will have any real interest in the details of your technical investigation. Instead, they want to know the answers to such questions as "Why was this investigation made? What is the significance of the problem it addresses? What am I supposed to do with the results of this investigation? What will it cost? What are the implications—for sales, for production, for the unions? What happens next? Who does it? Who is responsible?"

It is precisely this audience concern for nontechnical questions that causes so much trouble for young practicing engineers. Professionally, much of what the engineer spends his time doing is, at most, of only marginal concern to many of his audiences. His audiences ask questions about things which perhaps never entered his thoughts during his own technical activities when he received the assignment, defined the problem, and performed his investigation. These questions, however, must enter into his considerations when he writes his report.

Having defined the communication routes for a report you now know what audiences you will have and what questions they will want answered. The final step in our method of audience analysis is to assign priorities to your audiences. Classify them in terms of how they will use your report. In order of their importance to you (not in terms of their proximity to you), classify your audiences by these three categories:

- *Primary audiences*—who make decisions or act on the basis of the information a report contains.

- *Secondary audiences*—who are affected by the decisions and actions.

- *Immediate audiences*—who route the report or transmit the information it contains.

The *primary audience* for a report consists of those persons who will make decisions or act on the basis of the information provided by the report. The report overall should be designed to meet the needs of these users. The primary audience can consist of one person who will act in an official capacity, or it can consist of several persons representing several offices using the report. The important point here is that the primary audience for a report can consist of persons from any ring on the egocentric organization chart. They may be distant or in close proximity to the writer. They may be his organizational superiors, inferiors, or equals. They are simply those readers for whom the report is primarily intended. They are the top priority users.

In theory at least, primary audiences act in terms of their organizational roles rather than as individuals with distinctive idiosyncracies, predilections, and values. Your audience analysis should indicate when these personal concerns are likely to override organizational concerns. A typical primary audience is the decision maker, but his actual decisions are often determined by the evaluations and recommendations of staff personnel. Thus the report whose primary audience is a decision maker with line responsibility actually has an audience of staff personnel. Another type of primary audience is the production superintendent, but again his actions are often contingent upon the reactions of others.

In addition, because the report enters into a system, in time both the line and staff personnel will change; roles rather than individuals provide continuity. For this reason, it is helpful to remember the words of one engineer when he said, "A complete change of personnel could occur over the lifetime of my report." The report remains in the file. The report writer must not assume that his primary audience will be familiar with the technical assignment. He must design the report so that it contains adequate information concerning the reasons for the assignment, details of the procedures used, the results of the investigation, and conclusions and recommendations. This information is needed so that any future component of his primary audience will be able to use the report confidently.

The *secondary audiences* for a report consist of those persons other than primary decision makers or users who are affected by the information the report transmits into the system. These are the people whose activities are affected when a primary audience makes a decision, such as when production supervision has to adjust to management decisions. They must respond appropriately when a primary audience acts, such as when personnel and labor relations have to accommodate production line changes. The report writer must not neglect the needs of his secondary audiences. In tracing out his communication routes, he will identify several secondary audiences. Analysis of their needs will reveal what additional information the report should contain. This information is often omitted by writers who do not classify their audiences sufficiently.

The *immediate audience* for a report are those persons who route the report or transmit the information it contains. It is essential for the report writer to identify his immediate audiences and not to confuse them with his primary audiences. The immediate audience might be the report writer's supervisor or another middle management person. Yet usually his role will be to transmit information rather than to use the information directly. An information system has numerous persons who transmit reports but who may not act upon the information or who may not be affected by the information in ways of concern to the report writers. Often, a report is addressed to the writer's supervisor, but except for an incidental memo report, the supervisor serves only to transmit and expedite the information flow throughout the organizational system.

A word of caution: at times the immediate audience is also part of the primary audience; at other times the immediate audience is part of the secondary audience. For each report you write you must distinguish those among your readers who will function as conduits to the primary audience.

As an example of these distinctions between categories of report audiences, consider how audiences identified on the egocentric organization chart, Figure 4, can be categorized. Assume that the chemical engineer writes a report on a particular process improvement he has designed. The immediate audience might be his Group Leader. Another would be Manager I, transmitting the report to Manager II. The primary audiences might be Manager II and the Department Head; they would ask a barrage of nontechnical questions similar to those we mentioned a moment ago. They will decide whether or not the organization will implement the improvement recommended by the writer. The Department Head also could be part of the secondary audience by asking questions relating to labor relations and union contracts. Other secondary audiences, each asking different questions of the report, could be:

The person in contact with the funding agency, who will be concerned with budget and contract implications.

The person in contact with subcontractors, determining how they are affected.

The Group Leader, whose activities will be changed.

The "hands on" chemist, whose production responsibilities will be affected.

The technicians, whose job descriptions will change.

In addition to the secondary audiences on the egocentric organization chart, the report will have other secondary audiences throughout the organization—technical service and development, for example, or perhaps waste treatment.

At some length we have been discussing a fairly detailed method for systematic audience analysis. The method may have seemed more complicated than it actually is. Reduced to its basic ingredients, the method requires you, first, to identify all the individuals who will read the report, second, to characterize them, and third, to classify them. The *Matrix for audience analysis*, Figure 6, is a convenient device for characterizing and classifying your readers once you have identified them. At a glance, the matrix reveals what information you have and what information you still need to generate. Above all, the matrix forces you to think systematically. If you are able to fill in a good deal of specific information in each cell (particularly in the first six cells), you have gone a long way towards seeing how the needs of your audiences will determine the design of your report.

Characteristics Types of audiences	Operational	Objective	Personal
Primary	①	④	⑦
Secondary	②	⑤	⑧
Immediate	③	⑥	⑨

Fig. 6. Matrix for audience analysis.

We have not introduced a systematic method for audience analysis with the expectation that it will make your communication task easy. We have introduced you to the problems you must account for when you design your reports—problems you otherwise might ignore. You should, at least, appreciate the complexity of a report audience. Thus, when you come to write a report, you are less likely to make false assumptions about your audience. To develop this attitude is perhaps as important as to acquire the specific information the analysis yields. On the basis of this attitude, you now are ready to determine the specific purpose of your report.

What
to
report

RICHARD W. DODGE

At the time he wrote this article, Richard W.
Dodge was the editor of *Westinghouse Engineer*.

*T*echnical reports *can* be a useful tool for management—not only as a source of general information, but, more importantly, as a valuable aid to decision making. But to be effective for these purposes, a technical report must be geared to the needs of management.

Considerable effort is being spent today to upgrade the effectiveness of the technical report. Much of this is directed toward improving the writing abilities of engineers and scientists; or toward systems of organization, or format, for reports. This effort has had some rewarding effects in producing better written reports.

But one basic factor in achieving better reports seems to have received comparatively little attention. This is the question of audience needs. Or, expressed another way, "*What does management want in reports?*" This is an extremely basic question, and yet it seems to have had less attention than have the mechanics of putting words on paper.

A recent study conducted at Westinghouse sheds considerable light on this subject. While the results are for one company, probably most of them would apply equally to many other companies or organizations.

The study was made by an independent consultant with considerable experience in the field of technical report writing. It consisted of interviews with Westinghouse men at every level of management, carefully selected to present an accurate cross section. The list of questions asked is shown in Table 1. The results were compiled and analyzed and from the report several conclusions are apparent. In addition, some suggestions for report writers follow as a natural consequence.

Reprinted from *Westinghouse Engineer*, 22 (July–September, 1962), by permission of the Westinghouse Electric Corporation.

Table 1. Questions asked of managers

---

1. What types of reports are submitted to you?
2. What do you look for *first* in the reports submitted to you?
3. What do you want from these reports?
4. To what depth do you want to follow any one particular idea?
5. At what level (how technical and how detailed) should the various reports be written?
6. What do you want emphasized in the reports submitted to you? (Facts, interpretations, recommendations, implications, etc.)
7. What types of decisions are you called upon to make or to participate in?
8. What type of information do you need in order to make these decisions?
9. What types of information do you receive that you don't want?
10. What types of information do you want but not receive?
11. How much of a typical or average report you receive is useful?
12. What types of reports do you write?
13. What do you think your boss wants in the reports you send him?
14. What percentage of the reports you receive do you think desirable or useful? (In kind or frequency.)
15. What percentage of the reports you write do you think desirable or useful? (In kind or frequency.)
16. What particular weaknesses have you found in reports?

What management looks for in engineering reports

When a manager reads a report, he looks for pertinent facts and competent opinions that will aid him in decision making. He wants to know right away whether he should read the report, route it, or skip it.

To determine this, he wants answers fast to some or all of the following questions:

- What's the report about and who wrote it?
- What does it contribute?
- What are the conclusions and recommendations?
- What are their importance and significance?
- What's the implication to the Company?
- What actions are suggested? Short range? Long range?
- Why? By whom? When? How?

The manager wants this information in brief, concise, and meaningful terms. He wants it at the beginning of the report and all in one piece.

For example, if a summary is to convey information efficiently, it should contain three kinds of facts:

1. What the report is about;
2. The significance and implications of the work; and
3. The action called for.

To give an intelligent idea of what the report is about, first of all the problem must be defined, then the objectives of the project set forth. Next, the reasons for doing the work must be given. Following this should come the conclusions. And finally, the recommendations.

Such summaries are informative and useful, and should be placed at the beginning of the report.

The kind of information a manager wants in a report is determined by his management responsibilities, but how he wants this information presented is determined largely by his reading habits. This study indicates that management report reading habits are surprisingly similar. Every manager interviewed said he read the *summary* or abstract; a bare majority said they read the *introduction* and *background* sections as well as the *conclusions* and *recommendations*; only a few managers read the *body* of the report or the *appendix* material.

The managers who read the *background* section, or the conclusions and recommendations, said they did so ". . . to gain a better perspective of the material being reported and to find an answer to the all-important question: What do we do next?" Those who read the *body* of the report gave one of the following reasons:

1. Especially interested in subject;
2. Deeply involved in the project;
3. Urgency of problem requires it;
4. Skeptical of conclusions drawn.

And those few managers who read the *appendix* material did so to evaluate further the work being reported. To the report writer, this can mean but one thing: If a report is to convey useful information efficiently, the structure must fit the manager's reading habits.

The frequency of reading chart in Fig. 1 suggests how a report should be structured if it is to be useful to management readers.

Subject matter interest

In addition to what facts a manager looks for in a report and how he reads reports, the study indicated that he is interested in five broad technological areas. These are:

1. Technical problems;
2. New projects and products;
3. Experiments and tests;
4. Materials and processes;
5. Field troubles.

Managers want to know a number of things about each of these areas. These are listed in Table 2. Each of the sets of questions can serve as an effective check list for report writers.

In addition to these subjects, a manager must also consider market factors and organization problems. Although these are not the primary concern of the engineer, he should furnish information to management whenever technical aspects provide special evidence or insight into the problem being considered. For example, here are some of the questions about marketing matters a manager will want answered:

- What are the chances for success?
- What are the possible rewards? Monetary? Technological?
- What are the possible risks? Monetary? Technological?

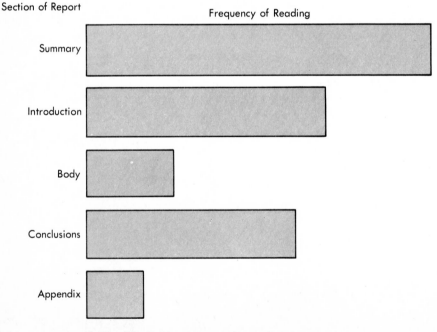

Fig. 1. How managers read reports

Table 2. What managers want to know

Problems
What is it?
Why undertaken?
Magnitude and importance?
What is being done? By whom?
Approaches used?
Thorough and complete?
Suggested solution? Best? Consider others?
What now?
Who does it?
Time factors?
New Projects and Products
Potential?
Risks?
Scope of application?
Commercial implications?
Competition?
Importance to Company?
More work to be done? Any problems?
Required manpower, facilities and equipment?
Relative importance to other projects or products?
Life of project or product line?
Effect on Westinghouse technical position?
Priorities required?
Proposed schedule?
Target date?

Tests and Experiments
What tested or investigated?
Why? How?
What did it show?
Better ways?
Conclusions? Recommendations?
Implications to Company?
Materials and Processes
Properties, characteristics, capabilities? Limitations?
Use requirements and environment?
Areas and scope of application?
Cost factors?
Availability and sources?
What else will do it?
Problems in using?
Significance of application to Company?
Field Troubles and Special Design Problems
Specific equipment involved?
What trouble developed? Any trouble history?
How much involved?
Responsibility? Others? Westinghouse?
What is needed?
Special requirements and environment?
Who does it? Time factors?
Most practical solution? Recommended action?
Suggested product design changes?

- Can we be competitive? Price? Delivery?
- Is there a market? Must one be developed?
- When will the product be available?

And, here are some of the questions about organization problems a manager must have answered before he can make a decision:

- Is it the type of work Westinghouse should do?
- What changes will be required? Organization? Manpower? Facilities? Equipment?
- Is it an expanding or contracting program?
- What suffers if we concentrate on this?

These are the kinds of questions Westinghouse management wants answered about projects in these five broad technological areas. The report writer should answer them whenever possible.

Level of presentation

Trite as it may sound, the technical and detail level at which a report should be written depends upon the reader and his use of the material. Most readers—certainly this is true for management readers—are interested in the significant material and in the general concepts that grow out of detail. Consequently, there is seldom real justification for a highly technical and detailed presentation.

Usually the management reader has an educational and experience background different from that of the writer. *Never* does the management reader have the same knowledge of and familiarity with the specific problem being reported that the writer has.

Therefore, the writer of a report for management should write at a technical level suitable for a reader whose educational and experience background is in a field different from his own. For example, if the report writer is an electrical engineer, he should write his reports for a person educated and trained in a field such as chemical engineering, or mechanical engineering, or metallurgical engineering.

All parts of the report *should preferably* be written on this basis. The highly technical, mathematical, and detailed material—if necessary at all—can and should be placed in the appendix.

Management responsibilities

The information presented thus far is primarily of interest to the report writer. In addition, however, management itself has definite responsibilities in the reporting process. These can be summed up as follows:

1. Define the project and the required reports;

2. Provide proper perspective for the project and the required reporting;

3. See that effective reports are submitted on time; and

4. See that the reports are properly distributed.

An engineering report, like any engineered product, has to be designed to fill a particular need and to achieve a particular purpose within a specific situation. Making sure that the writer knows what his report is to do, how it is to be used, and who is going to use it—all these things are the responsibilities of management. Purpose, use, and reader are the design factors in communications, and unless the writer knows these things, he is in no position to design an effective instrument of communication—be it a report, a memorandum, or what have you.

Four conferences at selected times can help a manager control the writing of those he supervises and will help him get the kind of reports he wants, when he wants them.

Step 1—At the beginning of the project. The purpose of this conference is to define the project, make sure the engineer involved knows what it is he's supposed to do, and specify the required reporting that is going to be expected of him as the project continues. What kind of decisions, for example, hinge upon his report? What is the relation of his work to the decision making process of management? These are the kinds of questions to clear up at this conference.

If the project is an involved one that could easily be misunderstood, the manager may want to check the effectiveness of the conference by asking the engineer to write a memorandum stating in his own words his understanding of the project, how he plans to handle it, and the reporting requirements. This can assure a mutual understanding of the project from the very outset.

Step 2—At the completion of the investigation. When the engineer has finished the project assignment—but before he has reported on it—the manager should have him come in and talk over the results of his work. What did he find out? What conclusions has he reached? What is the main supporting evidence for these conclusions? What recommendations does he make? Should any future action be suggested? What is the value of the work to the Company?

The broader perspective of the manager, plus his extensive knowledge of the Company and its activities, puts him in the position of being able to give the engineer a much better picture of the value and implications of the project.

The mechanism for getting into the report the kind of information needed for decision making is a relatively simple one. As the manager goes over the material with the engineer, he picks out points that need to be emphasized, and those that can be left out. This is a formative process that aids the engineer in the selection of material and evidence to support his material.

Knowing in advance that he has a review session with his supervisor, the chances are the engineer will do some thinking beforehand about the project and the results. Consequently, he will have formed some opinions about the significance of the work, and will, therefore, make a more coherent and intelligent presentation of the project and the results of his investigation.

This review will do something for the manager, too. The material will give him an insight into the value of the work that will enable him to converse intelligently and convincingly about the project to others. Such a preview may, therefore, expedite decisions influencing the project in one way or another.

Step 3—After the report is outlined. The manager should schedule a third conference after the report is outlined. At this session, the manager and the author should review the report outline step-by-step. If the manager is satisfied with the outline, he should tell the author so and tell him to proceed with the report.

If, however, the manager is not satisfied with the outline and believes it will have to be reorganized before the kind of report wanted can be written, he must make this fact known to the author. One way he can do this is to have the author tell him why the outline is structured the way it is. This usually discloses the organizational weakness to the author and consequently he will be the one to suggest a change. This, of course, is the ideal situation. However, if the indirect approach doesn't work, the more direct approach must be used.

Regardless of the method used to develop a satisfactory outline, the thing the manager must keep in mind is this: It's much easier to win the author's consent to structural changes at the outline stage, i.e., *before* the report is written, than afterward. Writing is a personal thing; therefore, when changes are suggested in organization or approach, these are all too frequently considered personal attacks and strained relations result.

Step 4—After the report is written. The fourth interview calls for a review and approval by the manager of the finished report and the preparation of a distribution list. During this review, the manager may find some sections of the report that need changing. While this is to be expected, he should limit the extent of these changes. The true test of any piece of writing is the clarity of the statement. If it's clear and does the job, the manager should leave it alone.

This four-step conference mechanism will save the manager valuable time, and, it will save the engineer valuable time. Also, it will insure meaningful and useful project reports—not an insignificant accomplishment itself. In addition, the process is an educative one. It places in the manager's hands another tool he can use to develop and broaden the viewpoint of the engineer. By eliminating misunderstanding and wasted effort, the review process creates a more helpful and effective working atmosphere. It acknowledges the professional status of the engineer and recognizes his importance as a member of the engineering department.

The writing of abstracts

CHRISTIAN K. ARNOLD

Christian K. Arnold was an instructor in English and a technical writer and editor before becoming Associate Executive Secretary of the Association of State Universities and Land-Grant Colleges.

*T*he most important section of your technical report or paper is the abstract. Some people will read your report from cover to cover; others will skim many parts, reading carefully only those parts that interest them for one reason or another; some will read only the introduction, results, and conclusions; but everyone who picks it up will read the abstract. In fact, the percentage of those who read beyond the abstract is probably related directly to the skill with which the abstract is written. The first significant impression of your report is formed in the readers' mind by the abstract; and the sympathy with which it is read, if it is read at all, is often determined by this first impression. Further, the people your organization wants most to impress with your report are the very people who will probably read no more than the abstract and certainly no more than the abstract, introduction, conclusions, and recommendations. And the people you should want most to read your paper are the ones for whose free time you have the most competition.

Despite its importance, you are apt to throw your abstract together as fast as possible. Its construction is the last step of an arduous job that you would rather have avoided in the first place. It's a real relief to be rid of the thing, and almost anything will satisfy you. But a little time spent in learning the "rules" that govern the construction of good abstracts and in practicing how to apply them will pay material dividends to both you and your organization.

The abstract—or summary, foreword, or whatever you call the initial thumbnail sketch of your report or paper—has two purposes: (1) it provides the specialist in the field with enough information about the report to permit him to decide whether he could read it with profit, and (2) it provides the administrator or executive with enough knowledge about what has been done

177

in the study or project and with what results to satisfy most of his administrative needs.

It might seem that the design specifications would depend upon the purpose for which the abstract is written. To satisfy the first purpose, for instance, the abstract needs only to give an accurate indication of the subject matter and scope of the study; but, to satisfy the second, the abstract must summarize the results and conclusions and give enough background information to make the results understandable. The abstract designed for the purpose can tolerate any technical language or symbolic shortcuts understood at large by the subject-matter group; the abstract designed for the second purpose should contain no terms not generally understood in a semitechnical community. The abstract for the first purpose is called a *descriptive abstract*; that for the second, an *informative abstract*.

The following abstract, prepared by a professional technical abstracter in the Library of Congress, clearly gives the subject-matter specialist all the help he needs to decide whether he should read the article it describes:

> Results are presented of a series of cold-room tests on a Dodge diesel engine to determine the effects on starting time of (1) fuel quantity delivered at cranking speed and (2) type of fuel-injection pump used. The tests were made at a temperature of $-10°F$ with engine and accessories chilled at $-10°F$ at least 8 hours before starting.

Regardless of however useful this abstract might be on a library card or in an index or an annotated bibliography, it does not give an executive enough information. Nor does it encourage everyone to read the article. If fact, this abstract is useless to everyone except the specialist looking for sources of information. The descriptive abstract, in other words, cannot satisfy the requirements of the informative abstract.

But is the reverse also true? Let's have a look at an informative abstract written for the same article:

> A series of tests was made to determine the effect on diesel-engine starting characteristics at low temperatures of (1) the amount of fuel injected and (2) the type of injection pump used. All tests were conducted in a cold room maintained at $-10°F$ on a commercial Dodge engine. The engine and all accessories were "cold-soaked" in the test chamber for at least 8 hours before each test. Best starting was obtained with 116 cu mm of fuel, 85 per cent more than that required for maximum power. Very poor starting was obtained with the lean setting of 34.7 cu mm. Tests with two different pumps indicated that, for best starting characteristics, the pump must deliver fuel evenly to all cylinders even at low cranking speeds so that each cylinder contributes its maximum potential power.

This abstract is not perfect. With just a few more words, for instance, the abstracter could have clarified the data about the amount of fuel delivered; do the figures give flow rates (what is the unit of time?) or total amount of fuel injected (over how long a period?)? He could easily have defined "best" starting. He could have been more specific about at least the more satisfactory type of pump: what is the type that delivers the fuel more evenly? Clarification of these points would not have increased the length of the abstract significantly.

The important point, however, is not the deficiencies of the illustration. In fact, it is almost impossible to find a perfect, or even near perfect, abstract, quite possibly because the abstract is the most difficult part of the report to write. This difficulty stems from the severe limitations imposed on its length, its importance to the over-all acceptance of the report or paper, and, with informative abstracts, the requirement for simplicity and general understandability.

The important point, rather, is that the informative abstract gives everything that is included in the descriptive one. The informative abstract, that is, satisfies not only its own purpose but also that of the descriptive abstract. Since values are obtained from the informative abstract that are not obtained from the descriptive, it is almost always worth while to take the extra time and effort necessary to produce a good informative abstract for your report or memo. Viewed from the standpoint of either the total time and effort expended on the writing job as a whole or the extra benefits that accrue to you and your organization, the additional effort is inconsequential.

It is impossible to lay down guidelines that will lead always to the construction of an effective abstract, simply because each reporting job, and consequently each abstract, is unique. However, general "rules" can be established that, if practiced conscientiously and applied intelligently, will eliminate most of the bugs from your abstracts.

1. *Your abstract must include enough* specific *information about the project or study to satisfy most of the administrative needs of a busy executive.* This means that the more important results, conclusions, and recommendations, together with enough additional information to make them understandable, must be included. This additional information will most certainly include an accurate statement of the problem and the limitations placed on it. It will probably include an interpretation of the results and the principal facts upon which the analysis was made, along with an indication of how they were obtained. Again, *specific* information must be given. One of the most common faults of abstracts in technical reports is that the information given is too general to be useful.

2. *Your abstract must be a self-contained unit, a complete report-in-miniature.* Sooner or later, most abstracts are separated from the parent report, and the abstract that cannot stand on its own feet independently must then either be rewritten or will fail to perform its job. And the rewriting, if it is done, will be done by someone not nearly as sympathetic with your study as you are. Even if it is not separated from the report, the

abstract must be written as a complete, independent unit if it is to be of the most help possible to the executive. This rule automatically eliminates the common deadwood phrases like "this report contains . . ." or "this is a report on . . ." that clutter up many abstracts. It also eliminates all references to sections, figures, tables, or anything else contained in the report proper.

3. *Your abstract must be short.* Length in an abstract defeats every purpose for which it is written. However, no one can tell you just how short it must be. Some authorities have attempted to establish arbitrary lengths, usually in terms of a certain percentage of the report, the figure given normally falling between three and ten per cent. Such artificial guides are unrealistic. The abstract for a 30-page report must necessarily be longer, percentagewise, than the abstract for a 300-page report, since there is certain basic information that must be given regardless of the length of the report. In addition, the information given in some reports can be summarized much more briefly than can that given in other reports of the same over-all dimensions. Definite advantages, psychological as well as material, are obtained if the abstract is short enough to be printed entirely on one page so that the reader doesn't even have to turn a page to get the total picture that the abstract provides. Certainly, it should be no longer than the interest span of an only mildly interested and very busy executive. About the best practical advice that can be given in a vacuum is to make your abstract as short as possible without cutting out essential information or doing violence to its accuracy. With practice, you might be surprised to learn how much information you can crowd into a few words. It helps, too, to learn to blue-pencil unessential information. It is perhaps important to document that "a meeting was held at the Bureau of Ordnance on Tuesday, October 3, 1961, at 2:30 P.M." somewhere, but such information is just excess baggage in your abstract: it helps neither the research worker looking for source material nor the administrator looking for a status or information summary. Someone is supposed to have once said, "I would have written a shorter letter if I had had more time." Take the time to make your abstracts shorter; the results are worth it. But be careful not to distort the facts in the condensing.

4. *Your abstract must be written in fluent, easy-to-read prose.* The odds are heavily against your reader's being an expert in the subject covered by your report or paper. In fact, the odds that he is an expert in your field are probably no greater than the odds that he has only a smattering of training in any technical or scientific discipline. And even if he were perfectly capable of following the most obscure, tortured technical jargon, he will appreciate your sparing him the necessity for doing it. T. O. Richards, head of the Laboratory Control Department, and R. A. Richardson, head of the Technical Data Department, both of the General Motors Corporation, have written that their experience shows the abstract cannot be made too elementary: "We never had [an abstract] . . . in which the explanations and terms were too simple." This requirement immediately eliminates the "telegraphic" writing often found in abstracts. Save footage by sound practices of economy and not by cutting out the articles and the transitional devices needed for smoothness and fluency. It also eliminates those obscure terms that you defend on the basis of "that's the way it's always said."

5. *Your abstract must be consistent in tone and emphases with the report proper, but it does not need to follow the arrangement, wording, or proportion of the original.* Data, information, and ideas introduced into the abstract must also appear in the report or paper. And they must appear with the same emphases. A conclusion or recommendation that is qualified in the report proper must not turn up without the qualification in the abstract. After all, someone might read both the abstract and the report. If this reader spots an inconsistency or is confused, you've lost a reader.

6. *Your abstract should make the widest possible use of abbreviations and numerals, but it must not contain any tables or illustrations.* Because of the space limitations imposed upon abstracts, the rules governing the use of abbreviations and numerals are relaxed for it. In fact, all figures except those standing at the beginning of sentences should be written as numerals, and all abbreviations generally accepted by such standard sources as the American Standards Association and "Webster's Dictionary" should be used.

By now you must surely see why the abstract is the toughest part of your report to write. A good abstract is well worth the time and effort necessary to write it and is one of the most important parts of your report. And abstract writing probably contributes more to the acquisition of sound expository skills than does any other prose discipline.

Graphs, illustrations, and tables

CHARLES T. BRUSAW, GERALD J. ALRED, AND WALTER E. OLIU

Charles T. Brusaw is a Senior Program Instructor for NCR Corporation. Gerald J. Alred is a member of the English Department at the University of Wisconsin-Milwaukee. Walter E. Oliu is a technical writer in the Division of Technical Writing and Document Control of the U.S. Nuclear Regulatory Commission. They are also the coauthors of *The Handbook of Technical Writing* (2nd ed., St. Martin's, 1982).

Graphs

A graph presents numerical data in visual form. This method has several advantages over presenting data in tables or within the text. Trends, movements, distributions, and cycles are more readily apparent in graphs than they are in tables. By providing a means for ready comparisons, a graph often shows a significance in the data not otherwise immediately apparent. Be aware, however, that although graphs present statistics in a more interesting and comprehensible form than tables do, they are less accurate. For this reason, they are often accompanied by tables giving the exact figures. There are many different kinds of graphs, most notably line graphs, bar graphs, pie graphs, and picture graphs.

Line graphs

The line graph, which is the most widely used of all graphs, shows the relationship between two sets of numbers by means of points plotted in relation to two axes drawn at right angles. The points, once plotted, are connected to one another to form a continuous line. In this way, what was merely a set of dots having abstract mathematical significance becomes graphic, and the relationship between the two sets of figures can easily be seen.

The line graph's vertical axis usually represents amounts, and its horizontal axis usually represents increments of time, as in Figure 1.

Line graphs with more than one line are common because they allow for comparisons between two sets of statistics for the same period of time. In creating such graphs, be certain to identify each line with a label or a legend, as shown in Figure 2. The difference between the two lines can be emphasized by shading the space between them.

Tips on preparing line graphs

1. Give the graph a title that describes the data clearly and concisely. Either center the title below the graph or align it with the left margin. If the title is too long to fit within the margins of the graph on one line, break the title into two or more lines of roughly equal length. In this case, the figure number should precede the first line of the title and the

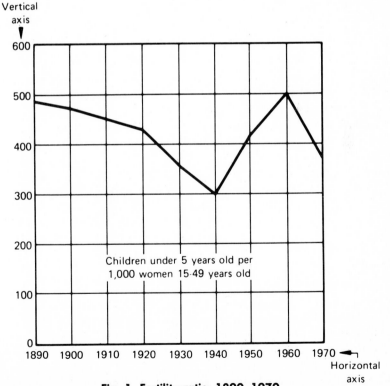

Fig. 1. Fertility ratio: 1890–1970

Source: U.S. Bureau of the Census

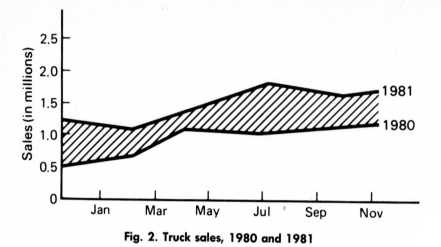

Fig. 2. Truck sales, 1980 and 1981

other lines should begin directly beneath the first word of the title. See Figure 12.

2. Assign a figure number if your report includes five or more illustrations. The figure number precedes the title, as in the examples throughout this entry.

3. Indicate the zero point of the graph (the point where the two axes meet). If the range of data shown makes it inconvenient to begin at zero, insert a break in the scale as in Figure 3.

4. Graduate the verticle axis in equal portions from the least amount at the bottom to the greatest amount at the top. Ordinarily, the caption for this scale is placed at the upper left.

5. Graduate the horizontal axis in equal units from left to right. If a caption is necessary, center it directly beneath the scale.

6. Graduate the vertical and horizontal scales so that they give an accurate visual impression of the data, since the angle at which the curved line rises and falls is determined by the scales of the two axes. The curve can be kept free of distortion if the scales maintain a constant ratio with each other. See Figures 4 and 5.

7. Hold grid lines to a minimum so that curved lines stand out. Since precise values are usually shown in a table of data accompanying a graph, detailed grid lines are unnecessary. Note the increasing clarity of the three graphs in Figures 6, 7, and 8.

8. Include a key (which lists and explains symbols) when necessary, as in Figure 7. At times a label will do just as well, as in Figure 8.

9. If the information comes from another source, include a source line below and on the same left margin as the caption, as in Figure 9.

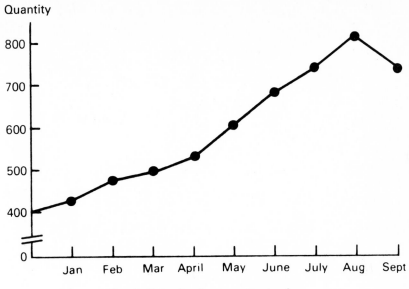

Fig. 3. Sales for January–September

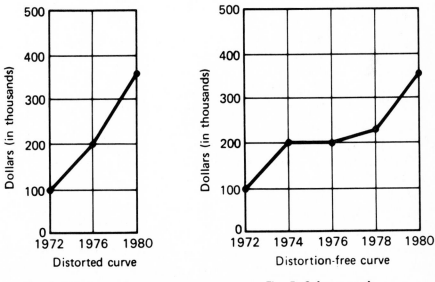

Fig. 4. Sales growth

Fig. 5. Sales growth

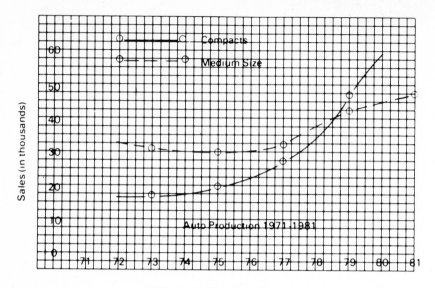

Fig. 6. Auto production

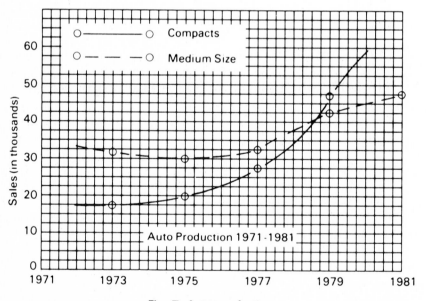

Fig. 7. Auto production

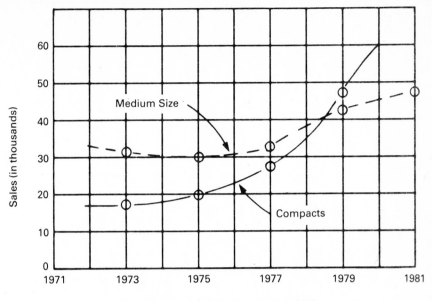

Fig. 8. Auto production, 1971–1981

10. Place explanatory footnotes below the graph, to the lower left.

11. Place all lettering horizontally if possible.

Bar graphs

Bar graphs consist of horizontal or vertical bars of equal width but scaled in length to represent some quantity. They are commonly used to show (1) quantities of the same item at different times, (2) quantities of different items for the same time period, or (3) quantities of the different parts of an item that make up the whole.

Figure 9 is an example of a bar graph showing varying quantities of the same item at the same time. Here each bar, which represents a different quantity of the same item, begins at the left scale. The left scale also provides additional information in the form of the percentage of the whole population each bar (and therefore each state) represents.

Some bar graphs show the quantities of different items for the same period of time. See Figure 10. (A bar graph with vertical bars is also called a column graph.)

Bar graphs can also show the different portions of an item that make up the whole. Here the bar is equivalent to 100 percent. It is then divided according to the appropriate proportions of the item sampled. This type of graph can be constructed vertically or horizontally and can indicate more than one whole where comparisons are necessary. See Figures 11 and 12.

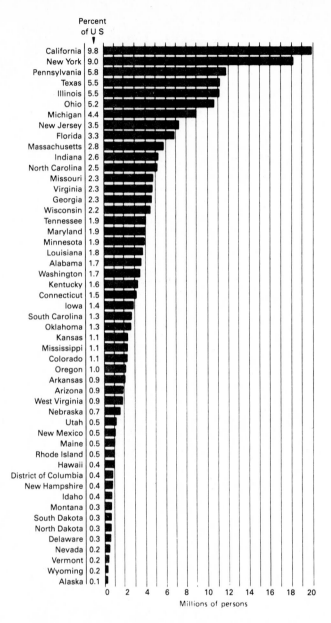

Fig. 9. States ranked by total population: 1970

Source: U.S. Bureau of the Census

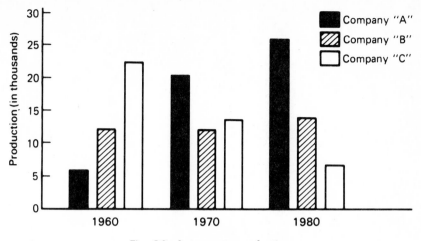

Fig. 10. Auto parts production

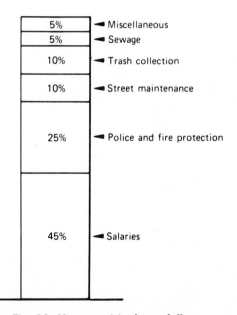

Fig. 11. Your municipal tax dollar

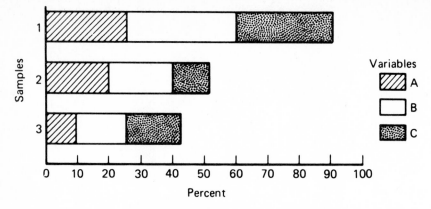

Fig. 12. Example of 100 percent bar graph showing proportions of three variables in three samples

If the bar is not labeled, each portion must be marked clearly by shading or crosshatching. Include a key that identifies the subdivisions.

Pie graphs

A pie graph presents data as wedge-shaped sections of a circle. The circle equals 100 percent, or the whole, of some quantity (a tax dollar, a bus fare, the hours of a working day), with the wedges representing the various ways in which the whole is divided. In Figure 13, for example, the circle stands for a city tax dollar, and it is divided into units equivalent to the percentage of the tax dollar spent on various city services.

Pie graphs provide a quicker way of presenting the same information that can be shown on a table; in fact, a table often accompanies a pie graph with a more detailed breakdown of the same information.

Tips on preparing pie graphs

1. The complete 360° circle is equivalent to 100 percent; therefore, each percentage point is equivalent to 3.6°.

2. To make the relative percentages clear, begin at the 12 o'clock position and sequence the wedges clockwise, from largest to smallest.

3. If you shade the wedges, do so clockwise and from light to dark.

4. Keep all labels horizontal and, most important, give the percentage value of each wedge.

5. Finally, check to see that all wedges, as well as the percentage values given for them, add up to 100 percent.

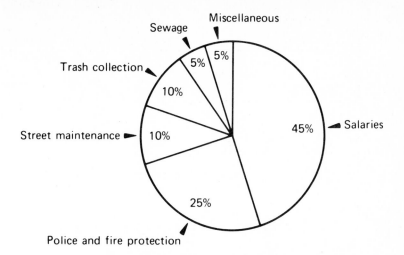

Fig. 13. Your municipal tax dollar

Although pie graphs have strong visual impact, they also have drawbacks. If more than five or six items are presented, the graph looks cluttered. Also, since they usually present percentages of something, they must often be accompanied by a table listing precise statistics. Further, unless percentages are shown on the individual sections, the reader cannot compare the values of the sections as accurately as is possible with a bar graph.

Picture graphs

Picture graphs are modified bar graphs that use picture symbols of the item presented. Each symbol corresponds to a specified quantity of the item. See Figure 14. Note that precise figures are included since the graph can present only approximate figures.

Tips on preparing picture graphs

    1. Make the symbol self-explanatory.

    2. Have each symbol represent a single unit.

    3. Show larger quantities by increasing the number of symbols rather than by creating a larger symbol. (It is difficult to judge relative sizes accurately.)

    4. Consult a good graphics manual for details about preparing graphs.

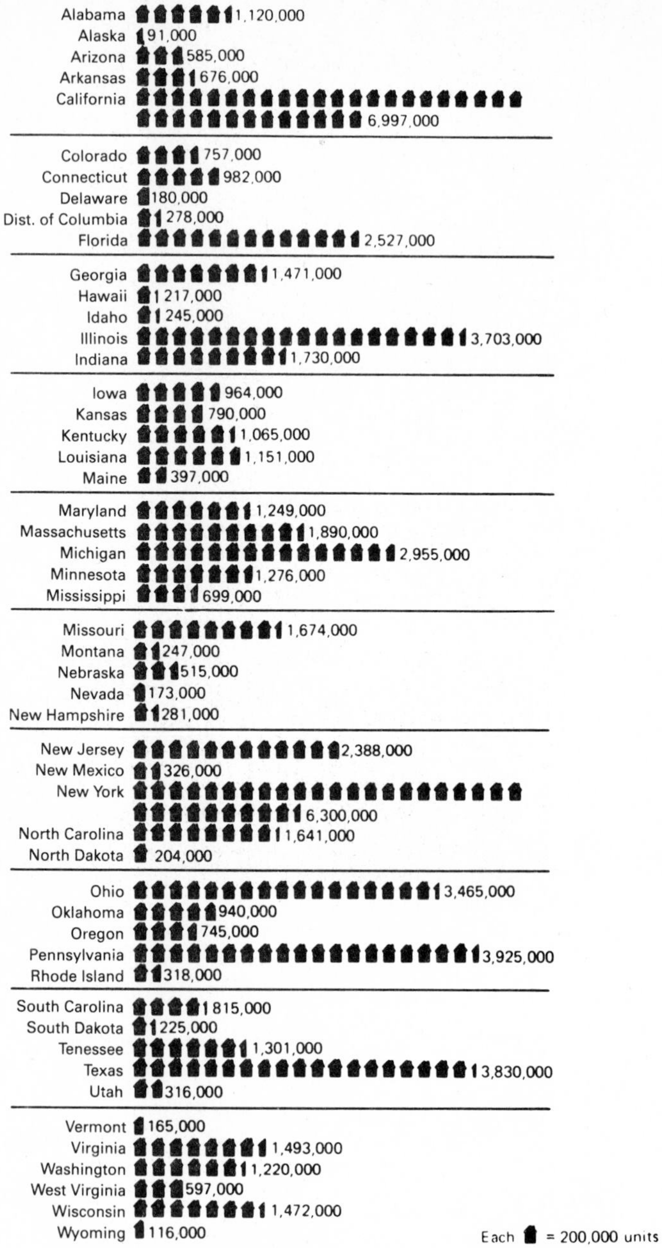

Alabama 🏠🏠🏠🏠🏠🏠1,120,000
Alaska 🏠91,000
Arizona 🏠🏠🏠585,000
Arkansas 🏠🏠🏠676,000
California 🏠🏠🏠🏠🏠🏠🏠🏠🏠🏠🏠🏠🏠🏠🏠🏠🏠🏠🏠🏠🏠🏠🏠🏠🏠🏠🏠🏠🏠🏠🏠🏠🏠🏠🏠 6,997,000

Colorado 🏠🏠🏠🏠757,000
Connecticut 🏠🏠🏠🏠🏠982,000
Delaware 🏠180,000
Dist. of Columbia 🏠🏠278,000
Florida 🏠🏠🏠🏠🏠🏠🏠🏠🏠🏠🏠🏠🏠2,527,000

Georgia 🏠🏠🏠🏠🏠🏠🏠1,471,000
Hawaii 🏠🏠217,000
Idaho 🏠🏠245,000
Illinois 🏠🏠🏠🏠🏠🏠🏠🏠🏠🏠🏠🏠🏠🏠🏠🏠🏠🏠3,703,000
Indiana 🏠🏠🏠🏠🏠🏠🏠🏠🏠1,730,000

Iowa 🏠🏠🏠🏠🏠964,000
Kansas 🏠🏠🏠🏠790,000
Kentucky 🏠🏠🏠🏠🏠🏠1,065,000
Louisiana 🏠🏠🏠🏠🏠🏠1,151,000
Maine 🏠🏠397,000

Maryland 🏠🏠🏠🏠🏠🏠🏠1,249,000
Massachusetts 🏠🏠🏠🏠🏠🏠🏠🏠🏠🏠1,890,000
Michigan 🏠🏠🏠🏠🏠🏠🏠🏠🏠🏠🏠🏠🏠🏠🏠2,955,000
Minnesota 🏠🏠🏠🏠🏠🏠🏠1,276,000
Mississippi 🏠🏠🏠🏠699,000

Missouri 🏠🏠🏠🏠🏠🏠🏠🏠1,674,000
Montana 🏠🏠247,000
Nebraska 🏠🏠🏠515,000
Nevada 🏠173,000
New Hampshire 🏠🏠281,000

New Jersey 🏠🏠🏠🏠🏠🏠🏠🏠🏠🏠🏠🏠2,388,000
New Mexico 🏠🏠326,000
New York 🏠🏠🏠🏠🏠🏠🏠🏠🏠🏠🏠🏠🏠🏠🏠🏠🏠🏠🏠🏠🏠🏠🏠🏠🏠🏠🏠🏠🏠🏠🏠🏠6,300,000
North Carolina 🏠🏠🏠🏠🏠🏠🏠🏠1,641,000
North Dakota 🏠204,000

Ohio 🏠🏠🏠🏠🏠🏠🏠🏠🏠🏠🏠🏠🏠🏠🏠🏠🏠3,465,000
Oklahoma 🏠🏠🏠🏠🏠940,000
Oregon 🏠🏠🏠🏠745,000
Pennsylvania 🏠🏠🏠🏠🏠🏠🏠🏠🏠🏠🏠🏠🏠🏠🏠🏠🏠🏠🏠3,925,000
Rhode Island 🏠🏠318,000

South Carolina 🏠🏠🏠🏠815,000
South Dakota 🏠🏠225,000
Tenessee 🏠🏠🏠🏠🏠🏠🏠1,301,000
Texas 🏠🏠🏠🏠🏠🏠🏠🏠🏠🏠🏠🏠🏠🏠🏠🏠🏠🏠🏠3,830,000
Utah 🏠🏠316,000

Vermont 🏠165,000
Virginia 🏠🏠🏠🏠🏠🏠🏠🏠1,493,000
Washington 🏠🏠🏠🏠🏠🏠🏠1,220,000
West Virginia 🏠🏠🏠597,000
Wisconsin 🏠🏠🏠🏠🏠🏠🏠1,472,000
Wyoming 🏠116,000

Each 🏠 = 200,000 units

Fig. 14. Number of housing units, by states: 1970

Source: U.S. Bureau of Census

Illustrations

The objective of using an illustration is to help your reader absorb the facts and ideas you are presenting. When used well, an illustration can convey an idea that words alone could never really make clear. Notice how the illustration in Figure 15 demonstrates how an accounting system can be computerized much more clearly and easily than could be done with words alone.

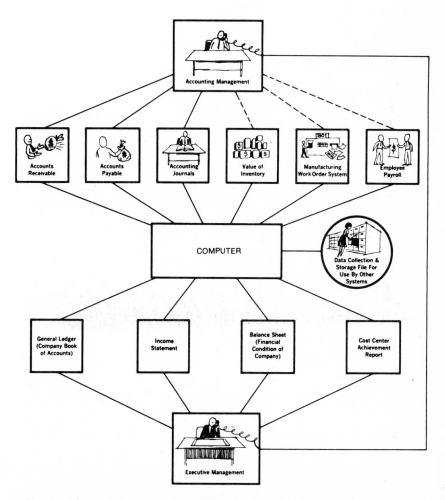

Fig. 15. The primary objective of the accounting information system is to remove the existing time lags in data reporting. Virtually every accounting function—billing, accounts receivable, accounts payable, tax liabilities, financial statements, and so forth—will be computerized and will contain built-in auditing and checking procedures.

Illustrations should never be used as ornaments, however; they should always be functional working parts of your writing. Be careful not to overillustrate; use an illustration only when you are sure that it makes a direct contribution to your reader's understanding of your subject. When creating illustrations, you must always consider your objective and your reader. You would use a different illustration of an X-ray machine, for example, for a high-school science class than you would for a group of medical doctors. Many of the attributes of good writing—simplicity, clarity, conciseness, directness—are equally important in creating and using illustrations.

The most common types of illustrations are photographs, graphs, tables, drawings, flowcharts, organizational charts, and maps. Your material will normally suggest one of these types when an illustration is needed.

Tips for creating and using illustrations

Each type of illustration has its unique strengths and weaknesses. . . . The guidelines presented here apply to most visual material you use to supplement the information in your text. Following these tips should help you create and present your visual material to good effect.

1. Keep the information as brief and simple as possible.

2. Try to present only one type of information in each illustration.

3. Label or caption each illustration carefully.

4. Include a key that identifies all symbols, when necessary.

5. Specify the proportions used or include a scale of relative distances, when appropriate.

6. Make the lettering horizontal for easy reading if possible.

7. Keep terminology consistent. Do not refer to something as a "proportion" in the text and as a "percentage" in the illustration.

8. Allow enough white space around and within the illustration.

9. Position the illustration as close as possible to the text that refers to it; however, an illustration should normally not appear ahead of the first text reference to it.

10. Be sure the significance of an illustration is clear in the text.

11. If figure numbers or table numbers are used, as they should be if five or more illustrations or tables are used, number the illustrations or tables consecutively.

12. If more than five illustrations or tables appear in a formal report, list them by title, together with figure and page numbers, under a separate

heading ("List of Figures" or "List of Tables") following the table of contents.

Presented with clarity and consistency, illustrations can help your reader focus on key portions of your report. Be aware, though, that even the best illustration only supplements the text. Your writing must carry the major burden of providing the context for the illustration and pointing out its significance.

Tables

A table is useful for showing large numbers of specific, related facts or statistics in a brief space. A table can present data in a more concise form than the text can, and a table is more accurate than graphic presentations because it provides numerous facts that a graph cannot convey. A table makes comparisons between figures easy because of the arrangement of the figures into rows and columns, although overall trends about the information are more easily seen in charts and graphs. See Table 1 for an example.

Guidelines for creating tables

Table Number. If you are using several tables, assign each a specific number; center the number and title above the table. The numbers are usually Arabic, and they should be assigned sequentially to the tables throughout the text. Tables should be referred to in the text by table number rather than by direction ("Table 4" rather than "the above table"). If there are more than five tables in your report or paper, give them a separate heading ("List of Tables") and list them by title and page number, together with any figure numbers, on a separate page immediately after the table of contents.

Title. The title, which is placed just above the table, should describe concisely what the table represents.

Boxhead. The boxhead carries the column headings. These should be kept concise but descriptive. Units of measurement, where necessary, should be specified either as part of the heading or enclosed in parentheses beneath the heading. Avoid vertical lettering where possible.

Stub. The left-hand vertical column of a table is the stub. It lists the items about which information is given in the body of the table.

Body. The body comprises the data below the boxhead and to the right of the stub. Within the body, columns should be arranged so that the terms to be compared appear in adjacent rows and columns. Leaders are sometimes used between figures to aid the eye in following data from column to column. Where no information exists for a specific item, leave an empty space to acknowledge the gap.

Caption →

Column captions →

Body

Table number

Boxhead

Stub

Rule

Footnote

Source line

→ Table 1 Recreational fresh-water angling by water-body type and geographical region*

GEOGRAPHICAL REGIONS	RESERVOIRS	MAN-MADE PONDS	NATURAL LAKES & PONDS	RIVERS & STREAMS	FARM PONDS
New England	130	40	570	410	410
Middle Atlantic	710	290	780	1200	630
East North Central	1200	760	3100	1600	1300
West North Central	810	550	1200	970	980
South Atlantic	1100	760	640	1500	1600
East South Central	890	630	190	670	1200
West South Central	1700	610	430	880	1300
Mountain	820	50	280	600	230
Pacific	950	200	820	1400	470
Totals	8300	3900	8000	9200	7800

* In thousands of anglers. Anglers who fished in more than one water body or region are represented in more than one category.

Source: U.S. Department of the Interior

Rules. These are the lines that separate the table into its various parts. Horizontal lines are placed below the title, below the body of the table, and between the column headings and the body of the table. They should not be closed at the sides. The columns within the table may be separated by vertical lines if they aid clarity.

Footnotes. Footnotes are used for explanations of individual items in the table. Symbols (*, #) or lower-case letters, rather than numbers, are ordinarily used to key table footnotes because numbers might be mistaken for the data in a numerical table.

Source Line. The source line, which identifies where the data were obtained, appears below any footnotes (when a source line is appropriate.)

Continued Lines. When a table must be divided so that it can be continued on another page, repeat the boxhead and give the table number at the head of each new page with a "continued" label ("Table 3, continued").

How to
lie with
statistics

DARRELL HUFF

Darrell Huff, a free-lance writer, expanded this
article into a book with the same title (Norton,
1954).

"The average Yaleman, Class of
'24," *Time* magazine reported last year after reading something in the New
York *Sun*, a newspaper published in those days, "makes $25,111 a year."

Well, good for him!

But, come to think of it, what does this improbably precise and salubrious
figure mean? Is it, as it appears to be, evidence that if you send your boy to
Yale you won't have to work in your old age and neither will he? Is this
average a mean or is it a median? What kind of sample is it based on? You
could lump one Texas oilman with two hundred hungry free-lance writers and
report *their* average income as $25,000-odd a year. The arithmetic is impecca-
ble, the figure is convincingly precise, and the amount of meaning there is in it
you could put in your eye.

In just such ways is the secret language of statistics, so appealing in a fact-
minded culture, being used to sensationalize, inflate, confuse, and oversim-
plify. Statistical terms are necessary in reporting the mass data of social and
economic trends, business conditions, "opinion" polls, this year's census. But
without writers who use the words with honesty and understanding and readers who know what they mean, the result can only be semantic nonsense.

In popular writing on scientific research, the abused statistic is almost
crowding out the picture of the white-jacketed hero laboring overtime without
time-and-a-half in an ill-lit laboratory. Like the "little dash of powder, little pot
of paint," statistics are making many an important fact "look like what she
ain't." Here are some of the ways it is done.

The sample with the built-in bias. Our Yale men—or Yalemen, as they
say in the Time-Life building—belong to this flourishing group.The exagger-

ated estimate of their income is not based on all members of the class nor on a random or representative sample of them. At least two interesting categories of 1924-model Yale men have been excluded.

First there are those whose present addresses are unknown to their classmates. Wouldn't you bet that these lost sheep are earning less than the boys from prominent families and the others who can be handily reached from a Wall Street office?

There are those who chucked the questionnaire into the nearest wastebasket. Maybe they didn't answer because they were not making enough money to brag about. Like the fellow who found a note clipped to his first pay check suggesting that he consider the amount of his salary confidential: "Don't worry," he told the boss. "I'm just as ashamed of it as you are."

Omitted from our sample then are just the two groups most likely to depress the average. The $25,111 figure is beginning to account for itself. It may indeed be a true figure for those of the Class of '24 whose addresses are known and who are willing to stand up and tell how much they earn. But even that requires a possibly dangerous assumption that the gentlemen are telling the truth.

To be dependable to any useful degree at all, a sampling study must use a representative sample (which can lead to trouble too) or a truly random one. If *all* the Class of '24 is included, that's all right. If every tenth name on a complete list is used, that is all right too, and so is drawing an adequate number of names out of a hat. The test is this: Does every name in the group have an equal chance to be in the sample?

You'll recall that ignoring this requirement was what produced the *Literary Digest's* famed fiasco.* When names for polling were taken only from telephone books and subscription lists, people who did not have telephones or *Literary Digest* subscriptions had no chance to be in the sample. They possibly did not mind this underprivilege a bit, but their absence was in the end very hard on the magazine that relied on the figures.

This leads to a moral: You can prove about anything you want to by letting your sample bias itself. As a consumer of statistical data—a reader, for example, of a news magazine—remember that no statistical conclusion can rise above the quality of the sample it is based upon. In the absence of information about the procedures behind it, you are not warranted in giving any credence at all to the result.

The truncated, or gee-whiz, graph. If you want to show some statistical information quickly and clearly, draw a picture of it. Graphic presentation is the thing today. If you don't mind misleading the hasty looker, or if you quite clearly *want* to deceive him, you can save some space by chopping the bottom off many kinds of graph.

---

* *Editor's note: The Literary Digest* predicted that Alfred Landon would defeat Franklin Roosevelt in the 1936 presidential election. Landon carried only two states.

Suppose you are showing the upward trend of national income month by month for a year. The total rise, as in one recent year, is 7 per cent. It looks like this:

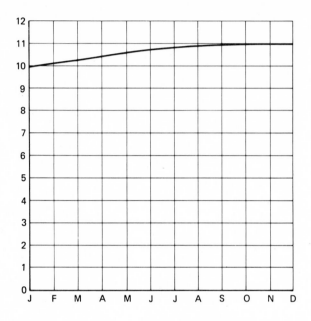

That is clear enough. Anybody can see that the trend is slightly upward. You are showing a 7 per cent increase and that is exactly what it looks like.

But it lacks schmaltz. So you chop off the bottom, this way:

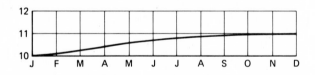

The figures are the same. It is the same graph and nothing has been falsified— except the impression that it gives. Anyone looking at it can just feel prosperity throbbing in the arteries of the country. It is a subtler equivalent of editing "National income rose 7 per cent" into ". . . climbed a whopping 7 per cent."

It is vastly more effective, however, because of that illusion of objectivity.

The souped-up graph. Sometimes truncating is not enough. The trifling rise in something or other still looks almost as insignificant as it is. You can make that 7 per cent look livelier than 100 per cent ordinarily does. Simply change the proportion between the ordinate and the abscissa. There's no rule against it, and it does give your graph a prettier shape.

But it exaggerates, to say the least, something awful:

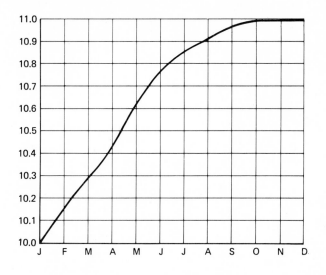

The well-chosen average. I live near a country neighborhood for which I can report an average income of $15,000. I could also report it as $3,500.

If I should want to sell real estate hereabouts to people having a high snobbery content, the first figure would be handy. The second figure, however, is the one to use in an argument against raising taxes, or the local bus fare.

Both are legitimate averages, legally arrived at. Yet it is obvious that at least one of them must be as misleading as an out-and-out lie. The $15,000-figure is a mean, the arithmetic average of the incomes of all the families in the community. The smaller figure is a median; it might be called the income of the average family in the group. It indicates that half the families have less than $3,500 a year and half have more.

Here is where some of the confusion about averages comes from. Many human characteristics have the grace to fall into what is called the "normal" distribution. If you draw a picture of it, you get a curve that is shaped like a bell. Mean and median fall at about the same point, so it doesn't make very much difference which you use.

But some things refuse to follow this neat curve. Income is one of them. Incomes for most large areas will range from under $1,000 a year to upward of $50,000. Almost everybody will be under $10,000, way over on the left-hand side of that curve.

One of the things that made the income figure for the "average Yaleman" meaningless is that we are not told whether it is a mean or a median. It is not that one type of average is invariably better than the other; it depends upon what you are talking about. But neither gives you any real information—and either may be highly misleading—unless you know which of those two kinds of average it is.

In the country neighborhood I mentioned, almost everyone has less than the average—the mean, that is—of $10,500. These people are all small farmers, except for a trio of millionaire week-enders who bring up the mean enormously.

You can be pretty sure that when an income average is given in the form of a mean nearly everybody has less than that.

The insignificant difference or the elusive error. Your two children Peter and Linda (we might as well give them modish names while we're about it) take intelligence tests. Peter's IQ, you learn, is 98 and Linda's is 101. Aha! Linda is your brighter child.

Is she? An intelligence test is, or purports to be, a sampling of intellect. An IQ, like other products of sampling, is a figure with a statistical error, which expresses the precision or reliability of the figure. The size of this probable error can be calculated. For their test the makers of the much-used Revised Stanford-Binet have found it to be about 3 per cent. So Peter's indicated IQ of 98 really means only that there is an even chance that it falls between 95 and 101. There is an equal probability that it falls somewhere else—below 95 or above 101. Similarly, Linda's has no better than a fifty-fifty chance of being within the fairly sizeable range of 98 to 104.

You can work out some comparisons from that. One is that there is rather better than one chance in four that Peter, with his lower IQ rating, is really at least three points smarter than Linda. A statistician doesn't like to consider a difference significant unless you can hand him odds a lot longer than that.

Ignoring the error in a sampling study leads to all kinds of silly conclusions. There are magazine editors to whom readership surveys are gospel; with a 40 per cent readership reported for one article and a 35 per cent for another, they demand more like the first. I've seen even smaller differences given tremendous weight, because statistics are a mystery and numbers are impressive. The same thing goes for market surveys and so-called public-opinion polls. The rule is that you cannot make a valid comparison between two such figures unless you know the deviations. And unless the difference between the figures is many times greater than the probable error of each, you have only a guess that the one appearing greater really is.

Otherwise you are like the man choosing a camp site from a report of mean temperature alone. One place in California with a mean annual temperature of 61 is San Nicolas Island on the south coast, where it always stays in the comfortable range between 47 and 87. Another with a mean of 61 is in the inland desert, where the thermometer hops around from 15 to 104. The deviation from the mean marks the difference, and you can freeze or roast if you ignore it.

The one-dimensional picture. Suppose you have just two or three figures to compare—say the average weekly wage of carpenters in the United States and another country. The sums might be $60 and $30. An ordinary bar chart makes the difference graphic.

That is an honest picture. It looks good for American carpenters, but perhaps it does not have quite the oomph you are after. Can't you make that difference appear overwhelming and at the same time give it what I am afraid is known as eye-appeal? Of course you can. Following tradition, you represent these sums by pictures of money bags. If the $30 bag is one inch high, you draw the $60 bag two inches high. That's in proportion, isn't it?

The catch is, of course, that the American's money bag, being twice as tall as that of the $30 man, covers an area on your page four times as great. And since your two-dimensional picture represents an object that would in fact have three dimensions, the money bags actually would differ much more than that. The volumes of any two similar solids vary as the cubes of their heights. If the unfortunate foreigner's bag holds $30 worth of dimes, the American's would hold not $60 but a neat $240.

You didn't say that, though, did you? And you can't be blamed, you're only doing it the way practically everybody else does.

The ever-impressive decimal. For a spurious air of precision that will lend all kinds of weight to the most disreputable statistics, consider the decimal.

Ask a hundred citizens how many hours they slept last night. Come out with a total of, say, 781.3. Your data are far from precise to begin with. Most people will miss their guess by fifteen minutes or more and some will recall five sleepless minutes as half a night of tossing insomnia.

But go ahead, do your arithmetic, announce that people sleep an average of 7.813 hours a night. You will sound as if you knew precisely what you are talking about. If you were foolish enough to say 7.8 (or "almost 8") hours it would sound like what it was—an approximation.

The semi-attached figure. If you can't prove what you want to prove, demonstrate something else and pretend that they are the same thing. In the daze that follows the collision of statistics with the human mind, hardly anybody will notice the difference. The semi-attached figure is a durable device guaranteed to stand you in good stead. It always has.

If you can't prove that your nostrum cures colds, publish a sworn laboratory report that the stuff killed 31,108 germs in a test tube in eleven seconds. There may be no connection at all between assorted germs in a test tube and the whatever-it-is that produces colds, but people aren't going to reason that sharply, especially while sniffling.

Maybe that one is too obvious and people are beginning to catch on. Here is a trickier version.

Let us say that in a period when race prejudice is growing it is to your advantage to "prove" otherwise. You will not find it a difficult assignment.

Ask that usual cross section of the population if they think Negroes have as good a chance as white people to get jobs. Ask again a few months later. As Princeton's Office of Public Opinion Research has found out, people who are most unsympathetic to Negroes are the ones most likely to answer yes to this question.

As prejudice increases in a country, the percentage of affirmative answers you will get to this question will become larger. What looks on the face of it like growing opportunity for Negroes actually is mounting prejudice and nothing else. You have achieved something rather remarkable: the worse things get, the better your survey makes them look.

The unwarranted assumption, or post hoc *rides again.* The interrelation of cause and effect, so often obscure anyway, can be most neatly hidden in statistical data.

Somebody once went to a good deal of trouble to find out if cigarette smokers make lower college grades than non-smokers. They did. This naturally pleased many people, and they made much of it.

The unwarranted assumption, of course, was that smoking had produced dull minds. It seemed vaguely reasonable on the face of it, so it was quite widely accepted. But it really proved nothing of the sort, any more than it proved that poor grades drive students to the solace of tobacco. Maybe the relationship worked in one direction, maybe in the other. And maybe all this is only an indication that the sociable sort of fellow who is likely to take his books less than seriously is also likely to sit around and smoke many cigarettes.

Permitting statistical treatment to befog casual relationships is little better than superstition. It is like the conviction among the people of the Hebrides that body lice produce good health. Observation over the centuries had taught them that people in good health had lice and sick people often did not. *Ergo,* lice made a man healthy. Everybody should have them.

Scantier evidence, treated statistically at the expense of common sense, has made many a medical fortune and many a medical article in magazines, including professional ones. More sophisticated observers finally got things straightened out in the Hebrides. As it turned out, almost everybody in those circles had lice most of the time. But when a man took a fever (quite possibly carried to him by those same lice) and his body became hot, the lice left.

Here you have cause and effect not only reversed, but intermingled.

There you have a primer in some ways to use statistics to deceive. A well-wrapped statistic is better than Hitler's "big lie": it misleads, yet it can't be pinned onto you.

Is this little list altogether too much like a manual for swindlers? Perhaps I can justify it in the manner of the retired burglar whose published reminiscences amounted to a graduate course in how to pick a lock and muffle a footfall: The crooks already know these tricks. Honest men must learn them in self-defense.

Strategies of persuasion

DAVID W. EWING

David W. Ewing is Executive Editor—Planning of
the *Harvard Business Review* and a member of
the faculty of Harvard Business School.

When we review reports, letters, and memoranda that get the intended results, we find a fascinating diversity of approaches. Some are gentle in approach, taking readers by the hand and leading them to a certain finding or recommendation. Others are brisk and abrupt. Some are objective in approach, carefully examining both sides of an idea, like a judge writing a difficult decision—at least until the end. Others burst with impatience to explain one side, and only one side, of a proposal or argument. Some flow swimmingly, others erupt like Mount Vesuvius.

How is it possible that communications taking such different approaches can all be effective? It is not sufficient to answer glibly, "It depends on the situation," or "Communications should mirror the personal style of the communicator." How does a written communication depend on the situation? Which style of the communicator should be reflected in a given communication? Examine the writing of most business executives, professionals, and public leaders, and you will find not one but *many* styles of exposition.

Rules every persuader should know

The explanation lies in a set of relationships among the communicator, the reader, the message, and the time-space environment. These relationships work in predictable ways and are an important part of the knowledge of every good business and professional writer. They come into play in the planning stages of writing, when the writer is considering how he or she will proceed, and in the main body of the presentation, to accomplish what he or she has promised in the opening paragraphs. The relationships affect some of the most important decisions a writer makes—the choice of ideas to use, the comparative emphasis to be given various arguments pro and con, the types of reasons and supporting material used, the establishment of credibility, and other matters.

It is convenient but self-defeating to follow fixed prescriptions for persuasion, such as to put your strongest arguments first or last or to identify with the readers. Such nostrums were fine for the age of patent medicines and snake-oil peddlers, but not for the age of diagnostic medicine. What is more, they are belittling. They assume that writers are witless. Good writers vary their approaches in response to their readings of different situations. Just as a good golfer plays an approach to a green differently depending on the wind, so a good writer uses different strategies depending on the crosscurrents of mood and feeling.

How should you choose your approach to a group of readers? What elements of the approach should be tailored to the situation? These are the topics of this chapter. Let us assume that the substance of the intended message is clear in your mind, and that you have written—or can write at any time—the kind of opening described in the previous chapter. . . .

1. Consider whether your views will make problems for readers.

J. C. Mathes and Dwight W. Stevenson tell of the student engineer who was asked to evaluate the efficiency of the employer plant's waste-treatment process.[1] He found that by making a simple change, the company could save more than $200,000 a year. Anticipating an enthusiastic response, he wrote up and delivered his report. Although he waited with great expectations, no accolades came. Why? What he hadn't counted on was that now his supervisors would have to explain to their bosses why they had allowed a waste of $200,000 a year. They were far from elated to read his report.

"If you want to make a man your enemy," Henry C. Link once said, "tell him simply, 'You are wrong.' This method works every time." Under the illusion that their sciences are "hard," physicists, biologists, and others may assume that it is necessary only to worry about setting the facts forth accurately. From posterity's standpoint, perhaps yes—but not from the standpoint of writing for current results. As Kenneth Boulding, head of the American Association for the Advancement of Science, has pointed out, the so-called "hard" sciences are in many ways "soft," and vice versa. "You can knock forever on a deaf man's door," said Zorba the Greek, and the deaf man can be a physicist as well as a marketing manager or public official.

If your views are bad news for readers, you proceed to report them, but with empathy and tact and an effort to put yourself in the reader's shoes. You work as carefully as if you were licking honey off a thorn.

2. Don't offer new ideas, directives, or recommendations for change until your readers are prepared for them.

"Should I state my surprising findings at the very beginning of my memorandum?" a writer asks. "Should I go slow with my heretical proposal and hold the reader's hand?" asks another.

---

[1] J. C. Mathes and Dwight W. Stevenson, *Designing Technical Reports* (Indianapolis: Bobbs-Merrill, 1976), pp. 18–19.

Generally speaking, the answer to all such questions depends on the extent of your audience's resistance to change, the amount of change you are asking for, the uncertainty in readers' minds as to your understanding of their situation, and what psychologists call the "perceived threat" of your communication, that is, how much it seems (to readers) to upset their values and interests. The more change, uncertainty, and/or threat, the slower you should proceed, the more carefully you should prepare your readers.

For instance, if your boss is enthusiastic about a new promotion scheme that he (or she) has paid a consultant $25,000 to devise, naturally you will want to go slow in shooting it down (at least if you want to stay in his good graces). In fact, any written criticism of the scheme is probably out of order until you have had a chance to talk with him and get a feel for the proper timing of any forthcoming criticism. When you do commit yourself to writing, you should probably review the arguments for the new scheme as fairly as possible, making it crystal clear to him that you understand them. Only then does it become timely to turn to the facts or conditions that, in your opinion, raise serious questions about the plan.

On the other hand, suppose the faulty promotion scheme is of little personal interest to your boss—it is not his (or her) "baby." Now the situation is different. You can launch right into the shortcomings, throwing your heaviest objections first. The fact is that the boss may not even want to know about the lesser objections, much less the supporting arguments once advanced for the plan; the main things he should know are that (a) the plan is in trouble and (b) the major, most compelling reasons why.

Clearly, this strategy is plain, everyday common sense—what you would normally do in communicating orally instead of in writing. Only in writing you must be more explicit and thorough, because a document lacks the expressiveness and visual advantages of a spoken dialogue.

Now consider another type of situation. Suppose it is your unhappy task to write a department manager that the extra appropriations he (or she) was promised have been cancelled. First of all, how would you handle it if you saw him often and could talk with him personally? . . . You would not (we hope) pussyfoot around trying to withhold the bad news from him. Also, once you had indicated the main message, you would probably backtrack a little and make it clear that the step is being taken with reluctance.

"Joe, it looks as if we're not going to be able to give you the extra budget we promised," you might say, getting down to business. That is the message in a nutshell—now for the review of common ground. "We know how much you have counted on getting those people and funds. There's no doubt you could manage them well and put them to good use. And we know that the morale of your people is involved in this, too. But the fact is that the sales we counted on are not coming in. We've got to cut somewhere, and, frankly, we feel it's got to be your department because . . . "

If you are communicating by letter or memorandum, the strategy is exactly the same (only "Joe" may now be "Mr. Wyncoop"). After your lead, you review the main needs as he and you understand them, perhaps spelling them out more than you would have in a face-to-face meeting but choosing the same ones. Only then do you turn to the new conditions that make it necessary to do an about-face.

3. Your credibility with readers affects your strategy.

In general, communication research indicates that the chances of opinion change vary with the communicator's authority with his or her readers. In their succinct summary of the field, *Persuasion,* social scientists Marvin Karlins and Herbert I. Abelson point out that credibility itself is a variable; that is, it can be influenced by the words of the communicator.[2] Above all, as psychologists repeatedly emphasize, credibility lies in the eye of the beholder.[3] . . .

In written communications there are two types of credibility. It may be given or it may be acquired. Between the two lies a world of practical difference.

Given credibility may result from your position in an organization. If, let us say, you are the boss writing directions to a subordinate, your credibility is likely to be high. Given credibility also may result from reputation—a well-known chemist has more credibility in communications about polymers than a good industrial engineer has, but the latter would possess more credibility in communications about time-and-motion studies. It may result from the individuals and groups the writer is associated with—if he or she is a member of the same trade union the reader belongs to, a union held in high esteem by both, he or she has more credibility in a memorandum on grievance procedures than a member of the board of directors would have.

Though you may be high in given credibility, you may yet need to remind some readers of the fact. In the case of an obvious credential, such as a position in an important organization, a letterhead may be enough to do the trick. Another device is to insert a few lines of biographical data at the head of a report or brochure. If you have had experience or associations that carry weight with the reader, perhaps you can interject them early in the message. "During a visit I had last week with Zach Jarvis," you might say, knowing that Dr. Zachary P. Jarvis is a magic name with your reader, or, "The Executive Committee of the Aberjona Basin Association asked me to join their meeting on Monday . . . " knowing that group carries a triple-A rating in the mind of the reader.

Of course, you do not want to overplay your hand at such name dropping. A report to a fairly diverse audience by a famous black organization began simply:

> The National Association for the Advancement of Colored People has for
> many years been dedicated to the task of defending the economic, social

[2] Marvin Karlins and Herbert I. Abelson. *Persuasion* (New York, Springer, 1970); see pp. 107–132.
[3] See, for example, Ralph L. Rosnow and Edward J. Robinson, Eds. *Experiments in Persuasion* (New York, Academic, 1967).

and political rights and interests of black Americans. The growing national debate about energy has led us to examine the question to ascertain the implications for black Americans.[4]

Again, an attorney of the American Civil Liberties Union, in a letter to members of the organization soliciting donations, began:

> My dear friend:
>
> I am the ACLU lawyer who went into court last April to defend freedom of speech in Skokie, Illinois, for a handful of people calling themselves "nazis."
>
> The case has had an enormous impact on my life.
>
> It has also gravely injured the ACLU financially. . . .[5]

Acquired credibility, on the other hand, is earned by thoughts and facts in the written message. I may not know you from Adam. Yet if you send me a letter or report that carefully, helpfully describes something I am interested in, you gain credibility in my estimation.

Some studies suggest that if you are low in given credibility and seek to acquire it with an audience, a useful technique is to cite ideas or evidence that support the reader's existing views.[6] As Disraeli once said, "My idea of an agreeable person is a person who agrees with me." The very fact that you feel confident and knowledgeable enough to articulate these views is likely to lift you several notches in the reader's estimation.

Still another approach is that old standby of persuaders—identifying yourself, in an early section, with the goals and interests of the audience. Possibly the most famous example of this strategy is the opening of Marc Antony's funeral oration in Shakespeare's *Julius Caesar:* "I come to bury Caesar, not to praise him. . . . "

Finally, you can acquire credibility by citing authorities who rate highly with your intended audience, or by exhibiting documentary evidence that, because of its source, lends prestige and authority to your proposals or ideas.

"For success in negotiation," says C. Northcote Parkinson and Nigel Rowe, "it is vitally important that people will believe what you say and assume that any promise you make will be kept. But it is no good saying: 'Trust me. Rely on my word.' Only politicians say that."[7]

Even if you have prestigious credentials, you cannot take too much for granted. In an age of television sets, radios, cassettes, and record players in every home, credibility—at least with the public—may come quicker for the

---

[4] *The Wall Street Journal*, January 12, 1978.
[5] David Goldberger, letter dated March 20, 1978.
[6] Karlins and Abelson, *op. cit.*, pp. 115–119.
[7] C. Northcote Parkinson and Nigel Rowe, "Better Communication: Business's Best Defense," *The McKinsey Quarterly*, Winter 1978, p. 26.

singer or comedian than for the judge, business executive, or medical researcher. In fact, because of an association with an organization or profession, you may be stereotyped as a member of "them" or "the establishment." It may behoove you to establish that you are a person with a name, a personality, certain interests, certain experiences—not just a nameless representative.

4. If your audience disagrees with your ideas or is uncertain about them, present both sides of the argument.

Behavioral scientists generally find that if an audience is friendly to a persuader, or has no contrary views on the topic and will get none in the near future, a one-sided presentation of a controversial question is most effective.[8] For instance, if your point is that sales of product X in the St. Louis territory could be doubled and you are writing to enthusiastic salespeople of product X, your best course is to concentrate on facts and examples showing the enormous potential of product X. There is no shortage of evidence showing that people generally prefer reading material that confirms their beliefs, and that they develop resistance to material that repudiates their beliefs. (As we shall see presently, however, this does not mean you cannot change their minds.)

But suppose your audience has not made up its mind, so far as you know? In this case you would do well to deal with *both* sides of the argument (or all sides, if there are more than two). Follow the same approach if the reader disagrees with you at the outset. For one thing, a two-sided presentation suggests to an uncertain or hostile audience that you possess objectivity. For another, it helps the reader remember your view by putting the pros and cons in relationship to one another. Also, it meets the reader's need to be treated as a mature, informed individual. As Karlins and Abelson point out:

> Conspicuously underlying your presentation is the assumption that the audience would be on your side if they only knew the truth. The other points of view should be presented with the attitude "it would be natural for you to have this idea if you don't know all the facts, but when you know all the facts, you will be convinced."[9]

Karlins and Abelson tested the reactions of audiences in post-war Germany to Voice of America broadcasts. They found that the most persuasive programs were those that included admissions of shortcomings in United States living conditions.[10]

Again, observation of businessmen's reactions to scores of *Harvard Business Review* articles advocating controversial measures convinces me that the

[8] Experiments supporting this conclusion are reported by Carl I. Hovland, Arthur A. Lumsdaine, and F. Sheffield in *Experiments on Mass Communication* (Princeton, Princeton University Press, 1949). Cited by Karlins and Abelson, *op. cit.*, p. 22.
[9] Karlins and Abelson, *op. cit.*, p. 26.
[10] See *Factors Affecting Credibility in Psychological Warfare Communications* (Washington, D.C., Human Resources Research Office, George Washington University, 1956).

most influential articles have been those that have acknowledged the short-comings, weaknesses, and limitations of their arguments. When an author wants to sell a new idea to a sophisticated audience, he or she should be candid about the soft spots in his argument.

5. Win respect by making your opinion or recommendation clear.

Although strategy may call for a two-sided argument, this does not mean you should be timid in setting forth your conclusions or proposals at the end. We assume here that you have definite views and seek to persuade your audience to adopt them. The two-sided approach is a *means* to that end; it does not imply compromising or obfuscating your conclusions. An official at Armour & Company once criticized many reports from subordinates to bosses on the ground that, after presenting much data, they concluded, in effect, "Here is what I found out and maybe we should do this or maybe we should do that." The typical response of a boss to such a memorandum, he noted, was to do nothing. Hence, the time taken both in writing and reading was wasted.[11]

6. Put your strongest points last if the audience is very interested in the argument, first if it is not so interested.

This question is referred to by social scientists as the "primacy-recency" issue in persuasion. The argument presented first is said to have primacy; the argument presented last, recency. Although studies of the question have produced inconsistent findings and no firm rules can be drawn, it appears that if your audience is deeply concerned with your subject you can afford to lead it along from the weakest points to the strongest. The audience's great interest will keep it reading, and putting the weaker points at the start tends to create rising reader expectations about what is coming. When you end with your strongest punch, therefore, you do not let readers down.

If your audience is not so concerned with the topic, on the other hand, it may be best to use the opposite approach. Now you cannot risk leading readers along a winding path. They may drop out before you reach the end. So grab their attention right at the beginning with your strongest argument or idea.

In any case, put the recommendation, facts, or arguments you most want the reader to *remember* first or last. Although experiments by social scientists on the primacy-recency issue are inconclusive, there is a firm pattern on the question of recall. The ideas you state first or last have a better chance of being remembered than the ideas stated in the middle of your appeal or case.

7. Don't count on changing attitudes by offering information alone.

"People are hostile to big business because they don't know enough facts about it," businesspeople are heard to say. Or, "If customers knew the truth

[11] John Ball and Cecil B. Williams, *Report Writing* (New York, Ronald, 1955).

about our costs, they would not object to our prices." Companies have poured large sums into advertising and public relations campaigns on this assumption; civic organizations have often based their hopes on it.

"The trouble with the assumption," states Karlins and Abelson, "is that it is almost never valid. There is a substantial body of research findings indicating that cognition—knowing something new—increasing information—is effective as an attitude change agent only under very specialized conditions."[12]

Social scientists do concede, however, that presentations of facts alone may strengthen the opinions of people who already agree with the persuader. The information reassures them and helps them defend themselves in discussions with others.

8. "Testimonials" are most likely to be persuasive if drawn from people with whom readers associate.

It is well known that a person's attitudes and opinions are strongly influenced by the groups to which he or she belongs or wants to belong—work units in a company, labor unions, bowling teams, social clubs, church associations, ethnic associations, and so on. To muster third-party support for your proposal or idea, therefore, you would do well to cite the behavior, findings, or beliefs of groups to which your readers belong. In so doing, you allay any feelings of isolation readers might have if tempted to follow your ideas. You suggest that they are not alone with you, that there is group support for the points being made.

As every school child learns, the predominant attitudes of a group toward individuals or regarding standards of behavior, performance, or status influence an individual member's perceptions. For instance, a study of boys at a camp demonstrated that their ratings of various individuals' performances at shooting and canoeing were biased by their knowledge of the status of the rated individuals in the camp society. Thus a boy generally regarded as a leader was seen as performing better with the rifle or canoe than was a boy generally regarded as a follower, even though the first boy's performance was not actually superior.[13]

In addition, it seems fair to say that as modern television, radio, records, and cassettes have brought national celebrities into the home and automobile, these people, too, have been stamped with approval or disapproval by millions of groups across the country.

Accordingly, if your readers are young, dissident, or "long hairs," refer to a Richard Dreyfus or a Joan Baez for supporting statements, not to a Gerald Ford or an Arnold Palmer. If your readers are electrical engineers, quote well-regarded scientific sources as your authority, not star salespeople or public relations people. Take into account also that the more deeply attached your readers are to a group, the greater the influence of the group norms on them.

[12] Karlins and Abelson, op. cit., p. 33.
[13] Ibid., p. 50.

For instance, one experiment by social scientists showed that the opinions of Catholic students who took their religion seriously were less influenced by the answers of nonserious Catholics than were the opinions of Catholic students who placed little value on their church membership.[14]

9. Be wary of using extreme or "sensational" claims and facts.

Both research in behavioral science and common sense confirm this rule.[15] Do not be misled by the fact that flashy journalists make successful use of extreme and bizarre cases to dramatize a story. The situation in business and professional writing is different from that in journalism.

When you seek the confidence and cooperation of your readers—and typically you do in the kinds of communications we deal with in this book—it is best to write in terms of the real world as you and they perceive it. Observable, believable, realistic statements carry more weight than any other kind. Although you want reader attention, you do not want to shock your audience with outlandish examples or arguments. These may help you to succeed in making the reader sit up—but they will also provoke distrust and suspicion.

Examples are common in the letters sections of newspapers. A writer who identified himself as a former vice-president of a well-known bank opposed a large power company's plan to build a new plant in a rural area near his town. His letter began as follows: "A great many of us . . . are profoundly disturbed by the proposal now being considered to disrupt and destroy the marvelous little valley southwest of [name of town], in order to build bigger and better power plants. This would be a devastating blow to the last unspoiled bit of country left in Connecticut. . . .[16]"

Like a batter who hits the first two pitches foul and quickly gets two strikes against him, this writer managed to distort the first two sentences he wrote. The proposed plant, though a very large one, would not "disrupt and destroy" the valley—only a small section of the valley area would be affected. Moreover, the valley was not "the last unspoiled bit of country" in the state— it was only a small parcel of the state's beautiful countryside. These exaggerations might have drawn cheers from rabid foes of the project, but the writer wasn't interested in appealing to them; he wanted to win uncommitted readers. At the very beginning, however, he antagonized them with hyperbole.

---

[14] See H. Kelley, "Salience of Membership and Resistance to Change of Group-Anchored Attitudes," *Human Relations*, August 1955, pp. 255–289. Cited in Karlins and Abelson, *op. cit.*, p. 58.

[15] See, for example, *Building Opposition to the Excess Profits Tax* (Princeton, Opinion Research Corporation, August 1952), and R. Weiss, "Conscious Technique for the Variation of Source Credibility," *Psychological Reports*, Vol. 20, 1969, p. 1159. Both cited in Karlins and Abelson, *op. cit.*, pp. 36–37.

[16] *Lakeville Journal*, April 2, 1970, p. 11.

10. **Tailor your presentation to the reasons for readers' attitudes, if you know them.**

Your chances of persuading readers are better if you can plan your appeal or argument to meet the main feelings, prejudices, or reasons for their beliefs. For instance, if reader beliefs are the result of their wanting to go along with certain groups they like or associate with, your best bet (as indicated earlier) is to show the acceptability of your point to these groups. If their attitudes reflect personal biases, such as an old grudge against someone in power, it is best to tailor your presentation to that prejudice. And so on.

Summarizing the implications of several behavioral studies. Karlins and Abelson present the example of three people who say they are against private ownership of industry. How should their reasons for this position influence one's choice of a strategy of persuasion? The authors explain:

> One of them feels that way because he has only been exposed to one side of the story and has nothing else on which to base his opinion. The way to change this man's opinion may be to expose him to facts, take him to visit some factories, meet some workers and supervisors. A second person is against private ownership because that is the prevailing norm or social climate in the circles in which he finds himself. His attitudes are caused by his being a part of a group and conforming to its standards. You cannot change this fellow just by showing him facts. The facts must be presented in an atmosphere which suggests a social reward for changing his opinion. Some kind of status appeal might be a start in that direction. A third person may have negative attitudes toward private industry because by making business the scapegoat for all his troubles, he can unload his pent-up feelings of bitterness and disappointment at the world for not giving him a better break. . . . Trying to change this third person with facts may actually do more harm than good. The more the evidence shows how wrong he is, the more he looks for reasons to support his beliefs. This kind of person can sometimes be influenced by helping him to understand why he has a particular attitude.[17]

11. **Never mention other people without considering their possible effect on the reader.**

Other people may, as we saw earlier, be introduced for the sake of "testimonials." More commonly, however, other people's names are mentioned in the course of explaining a situation, narrating an event, or completing the format of a message. This use of names, too, may affect the power of your message.

A reference to the actions of another person—however simple and unobtrusive it may seem to you the writer—may alter your relationship with readers. If readers consider that person a friend or enemy, their natural reaction is to begin thinking of the possible bearing of your communication on their

---

[17] Karlins and Abelson, *op. cit.*, p. 92.

friendship or antagonism. This reaction can have significant implications for your approach.

To illustrate, a doctoral student who had failed to meet his school's program requirements tried to muster faculty opinion in support of his petition for readmission by appearing daily at the entrance to the dining hall and handing out leaflets to faculty members. One such leaflet contained these words: "I am very unhappy about the strain my case has created for Professor [name of the program director]. I am distressed if last Friday's handout . . . created the impression that I was harping on his mistakes. I have told him and I tell you that I could understand his actions and decisions. . . ." The leaflet went on at some length to explain the doctoral student's feelings about the problem.

What this writer did not realize was the impact of the professor's name on his communication strategy. Almost everyone who received the leaflet was a colleague of the professor in question. Therefore the leaflet made it necessary for them to think of their relationship with the professor when they made up their minds about the petition. And their relationship with the professor was more important to them than their relationship with the doctoral student.

If the doctoral student considered it essential to mention the professor, he could have elected to: (1) try to win readers over while convincing them that their relations with the professor would not be affected, or (2) show that the professor was so far off base that readers were morally bound to risk their relationship with him. In the latter case, the leaflet should have contained ready-to-use arguments that readers could draw on in explaining to the professor why they sympathized with the doctoral student. Since the leaflet did neither of these things, it was a failure in persuasion.

Don't overlook the possible effect of distribution. Letters often go to third parties, with "cc" typed at the bottom followed by the names of those people. A memorandum often contains the names of several addressees in the "To" line at the top. Covering letters with reports may indicate several groups of readers. All this may affect your strategy. The background information that you could omit if writing only to Jones may be quite necessary if Brown, too, is an important reader; and the rather offhand treatment you give to a certain test or episode if writing to Jones and Brown might not be fitting at all if Larabee also is an intended reader. Many times the wise manager or professional rewrites part of a letter or memo after deciding to send a copy of it to an additional person who was not considered when the first draft was made.

Many people have strong feelings about "blind copies," that is, copies sent to persons other than those indicated after "cc" at the end of a letter or in the "To" line of a memorandum. Some people feel that blind copies never should be sent. Others feel that since a letter or memo is the property of the writer, he or she can distribute it at will. Although the latter view is legally correct, only an obtuse writer will distribute copies thoughtlessly if the content is in any way confidential, personal, or politically sensitive.

Sizing up your readers

We have a tendency to abstract written communications from real life, to act as if the customary ground rules of influence and persuasion don't apply to a message that is in writing. We act with a naiveté almost unheard of in our face-to-face relationships. Not seeing readers, we act as if they weren't real people. "If we write the information clearly, accurately, and correctly," we think wishfully to ourselves, "surely that satisfies the requirements of a piece of paper." But Josh Billings' puckish maxim, "As scarce as truth is, the supply has always been in excess of the demand," applies to truth on paper as well as truth in conversation.

Think of your intended readers as the real people they will be when they take your letter or report out of the "in-box." Only then can you decide intelligently what information and ideas to emphasize and in what order to present them. To help you think of readers as three-dimensional people, ask yourself some questions about their situation and relationships with you. Are they:

- Deeply or only mildly interested in the subject of your communication?

- Familiar or unfamiliar with your views, competence, and feelings about them?

- Knowledgeable or ignorant of your authority in the area discussed, your status, and your associations of possible importance to them?

- Committed or uncommitted to a viewpoint, opinion, or course of action other than the one you favor in your letter, report, or other document?

- Likely or unlikely to find your proposal, idea, finding, or conclusion threatening or requiring considerable change in their thought or behavior?

- Inclined or uninclined to think and feel the way they do about the subject because of identifiable reasons, prejudices, or experiences?

- Associated formally or informally with groups or organizations involved in some way with the idea or proposal you deal with?

With answers to questions like these in mind, you will not see your readers as shadows on the wall. They will sit across from you. You can write as if talking *with* them, not talking to them.

How to write effective proposals

LASSOR A. BLUMENTHAL

Lassor A. Blumenthal is the author of a number of books on business writing.

*V*irtually without exception, all proposals are like sales talks: they are an attempt to persuade someone to buy something or do something. In this sense, then, they are also like sales letters. . . .

We can divide proposals into two kinds: those for which there is fairly rigid format prescribed by the people to whom you're making the proposal, and those for which there is no rigidly prescribed format.

When the format is prescribed

When the format is prescribed, it's obviously a good idea to study the requirements closely. *Pay particularly close attention to the technical requirements.* That includes the following points:

1. *Number of copies:* Do they ask for one copy or for more than one?

2. *Typographic format:* Do they specify the page size? Whether to single- or double-space? How wide the margins should be? Whether to write on one or both sides of the page?

3. *Topic headings:* If they suggest that you cover certain specific topics, be sure that those topics are headlined clearly in your proposal.

4. *Length:* If they suggest minimum or maximum lengths to parts of the proposal, or to the whole thing, be sure to observe them.

When the format is not prescribed

When the format is not prescribed, your major task will be to make your report look good and be persuasive. These two requirements are quite distinct and are equally important. If your proposal looks bad, its persuasiveness will be diminished; if it's persuasive and it looks good, it will stand a better chance than an equally persuasive but poor-looking report.

Because appearance is so important, here are some suggestions to consider:

1. Use a good-looking cover. The more substantial your proposal, the heavier your cover should be. At the very least, use good quality letterhead paper. It will be more impressive if you use a simple cover. Check your stationery store: attractive, clear plastic bindings are available for less than a dollar. Thin cardboard bindings can also be bought for about the same price.

2. Use wide margins—at least 1½ inches on both sides.

3. Leave adequate margins at the top and bottom of the page—at least an inch.

4. Type on only one side of the sheet.

5. Double space throughout.

6. Start with a fresh sheet when you start a new section.

7. Use good quality white paper. Ask your stationery store for 20-pound rag bond—this has a good "feel." You might sample other papers which feel even heavier: they can help make even a mediocre report seem weightier.

8. Your proposal will undoubtedly be typewritten. No errors should show. Use correction fluid, chalk correction paper, or an eraser. And make sure your typewriter type is clean, so that the characters are sharp and clear. Your typewriter ribbon should be reasonably fresh, so that your characters are black and easily readable.

9. If possible, get copies of other proposals previously submitted to the place where you're submitting them. This will give you an idea of what's expected.

How to organize your proposal

The left-hand column below gives general principles for organizing each section of a proposal. The right-hand column shows how the proposal itself might be worded.

Title: The title should clearly state what the proposal is about.

A Proposal for an Improved System for Training Computer Programmers

State the purpose of the proposal. It can be as brief as a sentence. Ideally, it should be no longer than one or two paragraphs.

This is a proposal to provide the Amer Corporation with a better system for training computer programmers. Its major difference from the present system will be that it uses outside contractors instead of the existing in-house staff.

Summarize the major benefits to be derived from the proposal.

Explain the major features of the plan.

The major benefits to be expected are: lower training cost, faster training, and decreased turnover.

Major Features of the System

The major features of the proposed new system are:

1. All new programmers will be trained by the outside contractors.

2. The contractors will bill the company a flat fee for each trainee.

3. The contractors will be responsible for holding refresher courses annually for all programmers.

Why This Proposal Is Needed

Over the past five years, the quality of our programmers has deteriorated steadily, while their salaries and fringe benefits have risen steadily.

Those in a position to understand the problem agree that the basic trouble lies with the staff trainers. We must eliminate them and find new trainers immediately if we expect to bring our programmers up to at least minimal standards.

The system we recommend in this proposal solves all of our major problems, and we are unanimously in favor of it.

Explain why this is an important idea. This section might mention the following facts: What's wrong with—or lacking in—the existing conditions? Why is this the right time to carry out your proposal? If you're suggesting a new committee or department or organization, are you sure it's necessary? This is the place to explain why. NOTE: If the people to whom you're making the proposal know little or nothing of what you're talking about, it may be useful first to give them the background. In that case, you would start your proposal with this section, "Why This Proposal is Needed."

Then, you would put in a sentence that leads smoothly into the present first paragraph of the proposal. The sentence might be: "Therefore, we propose that the Amer Corp. adopt a new system for training computer programmers."

If your proposal is going to involve new people, or existing people in a new way, explain who they are and why they're the best ones for the job.

Who Will be Involved in the New System

To administer the new system, it's suggested that a committee of three be established. They are: Bill Almly, vice-president of computer operations; Jack Jones, chief auditor; and Harriet Freedberg, personnel director.

If you think it will be helpful, include resumes of the people you describe in this section. The resumes should be attached to the end of the report, as an appendix.

If your proposal is going to involve savings or expenditures, describe them in a separate section. If your financial figures are very detailed, it may be a good idea to simply summarize them here and to include a detailed analysis in an appendix.

The final paragraph is a call to action: it tells the reader what you want him or her to do or what decision you'd like to be forthcoming. Putting a deadline on the decision date may also help you to get an answer.

We suggest these three because:

1. They are most directly concerned with upgrading the efficiency of the computer personnel.

2. They have already been working on the problem and are familiar with its nuances.

3. Their departments are most directly affected by the poor quality of our existing training. . . .

What the New System Will Cost

Overall, we anticipate a saving of 10% to 20% at the end of the first year of operation under the proposed system. . . . The proposed revisions will cost the corporation annually about $20,000 compared to the $18,000 now being invested in programmer training. However, this increase out-of-pocket expense will be more than offset by a decrease in errors and in turnover.

Where Do We Go from Here?

As soon as management approves this idea, the new Training Committee will go into action. They will plan on filing monthly progress reports with top management.

We would like to have a management decision by April 15, so that we can implement our plans with minimal delay.

Other points to think about

The people who read your proposal are probably going to be a little frightened by it because it suggests changing things from the way they've been done in the past. This, of course, is a completely natural reaction: no matter how much we may want new suggestions and ideas, we're often leery of them. So, if you're writing a proposal, what can you do to allay the doubts in the reader's mind?

One way is to ask yourself: if I were in the reader's shoes, what kinds of questions would I have? If you make up a list of these questions and provide answers in your proposal, you'll have increased the chances of acceptance.

Here's a discussion of some of the questions which the reader is likely to have.

1. *How competent are the people making the proposal?*

If the person to whom the proposal is being made knows the proposal writers, there's not much of a problem. But often, the reader doesn't know the proposer. So, it may be useful to give information about the background or the experience of the people involved in the proposal. . . .

Consider whether additional information might be helpful. For example, if you're writing a proposal to another company, it may be helpful to include some letters from others praising you for past jobs. You might include these letters in an appendix, and you might refer to them near the beginning of your proposal—perhaps when you discuss who will be involved in carrying out the proposal.

2. *How practical and realistic is the proposal?*

Of course, the whole purpose of your proposal is to prove that what you're suggesting *is* practical. Nevertheless, the reader will have a number of doubts—especially if your idea is completely new. Here are some of the doubts likely to be present, and some suggestions about how to overcome them:

Is this the right time to do what the proposal suggests? If your reader is likely to ask this question, then you might answer it directly by including in your proposal a section titled: "Why the Time Is Right." In this section you can explain why it should be done now and what the consequences will be if it is not done now.

Is this proposal the right solution for the problem? Most business problems have several "right" solutions. If your reader is likely to ask whether a different proposal might be better, then you probably should deal with this in your proposal.

You may need to offer more explanations than are given in the sample proposal above. One way to handle this is with a section titled: "Some Solutions to the Problem." (In our sample proposal, we would put this section near the front, right after the "Major Benefits" section.)

A simple way of organizing this section is to discuss each of the alternatives briefly, then describe its advantages and disadvantages. From your description, it should be clear that the proposal you're suggesting is the best one.

Here's how such a section might be written for the sample proposal above:

Some solutions to the problem

The approach recommended in this proposal is one of several which we have considered. Those which seemed most sensible to us were:

1. Substantially upgrading the quality of the in-house training staff. The advantage of this approach is that it would enable us to continue the training system we now have, but in a more effective manner. The disadvantage is that it would be quite costly: we estimated an increase of between $10,000 and $15,000 annually.

2. Contracting all company-operated computer services to an outside firm. The advantage of this approach is that it eliminates entirely the problem of training our own people. The disadvantages are several: we would lose direct control over our computer operations; we would have to drastically write down a considerable amount of expensive equipment, and the costs would be 20% to 30% higher than they now are.

3. A middle course between 1 and 2: keeping our in-house staff but contracting out the training. This is the course we are recommending in this proposal.

Are the people recommending it enthusiastic about it? A proposal that expresses enthusiasm engenders confidence. It's very much like a salesperson who's enthusiastic about the worth of a product: you tend to trust it and believe in it more than if it's put forward with indifference. In our sample proposal, this enthusiasm is suggested in a number of ways: it's in the title, which talks about an "improved system." It's in the first sentence, which mentions a "better system." It's expressed in the last sentence of the section "Why This Proposal Is Needed": "The system we recommend . . . solves all of our major problems, and we are all on record as being enthusiastically in favor of it."

Does the proposal suggest adequate staff and facilities for the job? If your proposal suggests setting up a new program that will involve a number of people and perhaps additional space or facilities, describe them in enough detail to show the reader that you've thought the problem through. The body of your proposal might simply describe the staffing in general terms, and you might go into considerable detail in an appendix.

3. *How important is the proposal to the company?*

The more important you can make the proposal seem to the well-being of the company, the more seriously it will be considered. In the sample proposal, this urgency is referred to in the section "Why This Proposal Is Needed." Among the questions a prospective reader will ask are:

Who'll benefit from the project—and in what way?

Will conditions improve measurably if it's successful?

Will we be worse off if it fails than if we had done nothing?

4. *Is it original and creative?*

Most people have mixed feelings about things that are new and creative. On the one hand, they like to be associated with creative projects because they're interesting and usually exciting, and they get talked about. On the other hand, they're risky: that which is new takes getting used to, and sometimes, it's hard to get used to new ideas without considerable discomfort. So, in writing your proposal, give some thought to the person who's going to make the final decision on whether to approve it. Is he or she likely to respond more favorably to something that's new and innovative or to something that's more conservative? Then, consider whether you can tilt your report accordingly.

Here are some examples of how you might slant a proposal in one direction or the other.

If you plan to suggest that your ideas are creative and original, you can introduce it with a paragraph like this:

> The problem we face is a severe one, and its solution demands bold and innovative thinking. Admittedly, it requires courage and persistence for any innovation to succeed. But we believe that our proposal is sufficiently creative to call forth the very best qualities of everyone who participates in it: once it's initiated, we are certain that it will be seen as natural, inevitable, and an immense improvement over the program it replaces.

On the other hand, if you want to suggest that your ideas, while creative and original, are also conservative and cautious, you might introduce them with a paragraph like this:

> Our idea adapts the best of the past to the new conditions we face today. Rather than abandon what has proved itself to be true and valuable, we have attempted to define its essence and give it new life and new meaning for the world we are moving into. For this reason, we think our proposal is eminently sound: it combines solid experience with an open-minded willingness to accept the future and to make the most of it. This approach has accounted for much of our success in the past, and we believe it is the correct approach for guiding us toward a successful future.

Some other points you might want to touch on in your proposal which relate to its originality and creativity:

a. Does it duplicate or overlap other existing or forthcoming programs?

b. Will it disturb or upset other existing or forthcoming programs?

5. *Is the budget soundly conceived?*

The soundness of your budget will be one of the most thoroughly scrutinized parts of your proposal. Ask yourself two questions:

a. Is the budget adequate for the job I want to accomplish? And, is it obviously not so lavish that it appears wasteful?

b. Does your budget include a contingency sum, in case expenses run higher than expected?

6. *Do you need to provide for evaluation of the project?*

It may help to make your proposal more persuasive if you include a section describing how you plan to evaluate the results of the project. Think about these things:

a. How often should it be evaluated? Every month? Every six months? Once a year?

b. What kinds of evaluation are needed: will you evaluate how well you've achieved specific objectives, how close you've come to general goals, how people feel about the project, or how the money has been spent?

c. Who'll do the evaluating? Will it be an independent person or committee, or will it be people who are deeply involved in the project? In either case, you should be able to present a couple of good reasons explaining why these are the best people to do the job.

Ten report writing pitfalls: how to avoid them

VINCENT VINCI

Vincent Vinci was Director of Public Relations for
Lockheed Electronics when he wrote this article.

*T*he advancement of science moves
on a pavement of communications. Chemists, electrical engineers, botanists,
geologists, atomic physicists, and other scientists are not only practitioners but
interpreters of science. As such, the justification, the recognition and the rewards within their fields result from their published materials.

Included in the vast field of communications is the report, a frequently
used medium for paving the way to understanding and action. The engineering
manager whose function is the direction of people and programs receives and
writes many reports in his career. And therefore the need for technical reports
that communicate effectively has been internationally recognized.

Since scientific writing is complicated by specialized terminology, a need
for precision and the field's leaping advancement, the author of an engineering
report can be overwhelmed by its contents. The proper handling of contents
and communication of a report's purpose can be enhanced if the writer can
avoid the following 10 pitfalls.

Pitfall 1: Ignoring your audience

In all the forms of communications, ignoring your audience in the preparation
of a report is perhaps the gravest transgression. Why? All other forms of
communication such as instruction manuals, speeches, books and brochures,
are directed to an indefinable or only partially definable audience. The report,

on the other hand, is usually directed to a specific person or group and has a specific purpose. So, it would certainly seem that if one knows both the "who" and the "why," then a report writer should not be trapped by this pitfall.

But it is not enough to know the who and why, you need to know "how." To get to the how, let's assume that the reader is your boss and has asked you to write a trip report. You are to visit several plants and report on capital equipment requirements. Before you write the first word you will have to find out what your boss already knows about these requirements. It is obvious that he wants a new assessment of the facilities' needs. But, was he unsatisfied with a recent assessment and wants another point of view, or is a new analysis required because the previous report is outdated—or does he feel that now is the time to make the investment in facilities so that production can be increased over the next five years? That's a lot of questions, but they define both the who and why of your trip and, more importantly, your report.

By this time you may get the feeling that I suggest you give him exactly what he wants to read. The answer is yes and no. No, I don't mean play up to your boss's likes and dislikes. I do mean, however, that you give him all the information he needs to make a decision—the pros and the cons.

I mean also that the information be presented in a way that he is acclimated to in making judgments. For example, usually, a production-oriented manager or executive (even the chief executive) will think in terms of his specialty. The president of a company who climbed the marketing ladder selling solvents will think better in marketing terms. Therefore, perhaps the marketing aspects of additional equipment and facilities should be stressed. You should also be aware that if you happen to be the finance director, your boss will expect to see cost/investment factors too.

A simple method for remembering, rather than ignoring, your audience is to place a sheet of paper in front of you when you start to write your report. On the paper have written in bold letters WHO, WHY and HOW, with the answers clearly and cogently defined. Keep it in front of you throughout the preparation of your report.

Pitfall 2: Writing to impress

Nothing turns a reader off faster than writing to impress. Very often reports written to leave a lasting scholarly impression on top management actually hinder communication.

Generally, when a word is used to impress, the report writer assumes that the reader either knows its meaning or will take the trouble to look it up. Don't assume that a word familiar to you is easily recognized by your reader. I recall a few years ago, there was a word "serendipity" which became a fashionable word to impress your reader with. And there was "fulsome," and "pejorative," and more. All are good words, but they're often misused or misapplied. They were shoved into reports to impress, completely disregarding the reader. Your objective is that your reader comprehend your thoughts, and there should be a minimum of impediments to understanding—understanding with first reading, and no deciphering.

Unfortunately, writing to impress is not merely restricted to use of obscure words but also includes unnecessary detail and technical trivia. Perhaps the scientist, chemist, chemical engineer and others become so intrigued with technical fine points that the meaningful (to your audience) elements of a report are buried. And quite often the fault is not so much a lack of removing the chaff from the grain but an attempt to technically impress the reader. Of course there exist reports that are full of technical detail because the nature of the communication is to impart a new chemical process, compound or technique. Even when writing this kind of report, you should eliminate any esoteric technical facts that do not contribute to communication, even though you may be tempted to include them to exhibit your degree of knowledge in the field.

Pitfall 3: Having more than one aim

A report is a missile targeted to hit a point or achieve a mission. It is not a barrage of shotgun pellets that scatter across a target indiscriminately.

Have you ever, while reading a report, wondered where or what it was leading to—and even when you finished you weren't quite sure? The writer probably had more than one aim, thereby preventing you from knowing where the report was heading.

Having more than one aim is usually the sign of a novice writer, but the pitfall can also trip up an experienced engineer if he does not organize the report toward one objective.

It is too easy to say that your report is being written to communicate, to a specific audience, information about your research, tests, visit, meeting, conference, field trip, progress or any other one of a range of activities that may be the subject. If you look at the first part of the sentence, you will see that "specific audience" and "information" are the key words that have to be modified to arrive at the goal of your report. For instance, you must define the specific audience such as the "members of the research council," "the finance committee" or "the chief process engineer and his staff."

Secondly, you need to characterize the information, such as "analysis of a new catalytic process," "new methods of atomic absorption testing" or "progress on waste treatment programs." You should be able to state the specific purpose of your report in one sentence: e.g., "The use of fibrous material improves scrubber efficiency and life—a report to the product improvement committee."

When you have arrived at such a definition of your purpose and audience, you can then focus both the test results and analysis toward that purpose, tempered with your readers in mind.

The usual error made in writing reports is to follow the chronology of the research in the body of the report with a summary of a set of conclusions and recommendations attached. The proper procedure to follow is to write (while focusing on your report goal) the analysis first (supported by test essentials or any other details), then your introduction or summary—sort of reverse chronology. But be sure that your goal and audience are clearly known, because they become the basis of organizing your report.

Pitfall 4: Being inconsistent

If you work for an international chemical firm, you may be well aware of problems in communicating with plant managers and engineers of foreign installations or branches. And I'm not referring to language barriers, because for the most part these hurdles are immediately recognized and taken care of. What is more significant is units of measure. This problem is becoming more apparent as the United States slowly decides whether or not to adopt the metric system. Until it is adopted, your best bet is to stick to one measurement system throughout the report. Preferably, the system chosen should be that familiar to your audience. If the audience is mixed you should use both systems with one (always the same one) in parentheses. Obviously, don't mix units of measure because you will confuse or annoy your readers.

Consistency is not limited to measurements but encompasses terms, equations, derivations, numbers, symbols, abbreviations, acronyms, hyphenation, capitalization and punctuation. In other words, consistency in the mechanics of style will avoid work for your reader and smooth his path toward understanding and appreciating the contents of the report.

If your company neither has a style guide nor follows the general trends of good editorial practice, perhaps you could suggest instituting a guide. In addition to the U.S. Government Printing Office Style Manual, many scientific and engineering societies have set up guides which could be used.

Pitfall 5: Overqualifying

Chemical engineers, astronomers, geologists, electrical engineers, and scientists of any other discipline have been educated and trained to be precise. As a result, they strive for precision, accuracy, and detail. That tends to work against the scientist when it comes to writing. Add to that the limited training received in the arts, and you realize why written expression does not come easily.

Most reports, therefore, have too many modifiers—adjectives, clauses, phrases, adverbs and other qualifiers. Consider some examples: the single-stage, isolated double-cooled refractory process breakdown, or, the angle of the single-rotor dc hysteresis motor rotor winding. To avoid such difficult-to-comprehend phrases, you could in the first example write "the breakdown of the process in single-stage, isolated double-cooled refractories," and in the second, "the angle of the rotor winding in single-rotor dc hysteresis motors can cause . . . ," and so on. This eliminates the string of modifiers and makes the phrase easier to understand.

Better still, if your report allows you to say at the beginning that the following descriptions are only related to "single-stage, isolated double-cooled refractories" or "single-rotor dc hysteresis motors," you can remove the cumbersome nomenclature entirely.

In short, to avoid obscuring facts and ideas, eliminate excessive modifiers. Try to state your idea or main point first and follow with your qualifying phrases.

Pitfall 6: Not defining

Dwell, lake, and barn, are all common words. Right? Right and wrong. Yes, they are common to the non-scientist. To the mechanical engineer, dwell is the period a cam follower stays at maximum lift; to a chemical engineer, lake is a dye compound; and to an atomic physicist, a barn is an atomic cross-sectional area (10^{-24} cm^2).

These three words indicate two points: first, common words are used in science with other than their common meaning; and second, terms need to be defined.

In defining terms you use in a report, you must consider what to define and how to define. Of the two, I consider what to define a more difficult task and suggest that you review carefully just which terms you need define. If you analyze the purpose, the scope, the direction and your audience (reader/user), you will probably get a good handle on such terms.

"How to define" ranges from the simple substitution of a common term for an uncommon one, to an extended or amplified explanation. But whatever the term, or method of definition, you need to slant it both to the reader and to the report purpose.

Pitfall 7: Misintroducing

Introductions, summaries, abstracts and forewords—whatever you use to lead your reader into your report, it should not read like an exposition of a table of contents. If it does, you might as well let your audience read the table of contents.

The introduction, which should be written after the body of the report, should state the subject, purpose, scope, and the plan of the report. In many cases, an introduction will include a summary of the findings or conclusions. If a report is a progress report, the introduction should relate the current report to previous reports. Introductions, then, not only tell the sequence or plan of the report, but tell the what, how and why of the subject as well.

Pitfall 8: Dazzling with data

Someone once said that a good painter not only knows what to put in a painting, but more importantly he knows what to leave out. It's much the same with report writing. If you dazzle your reader with tons of data, he may be moved by the weight of the report but may get no more out of it than that.

The usual error occurs in supportive material that many engineers and scientists feel is necessary to give a report scientific importance. The truth is that successful scientific writing (which includes reports) is heavily grounded in reality, simplicity, and understanding—not quantity.

The simplest way to evaluate the relevancy of information is to ask yourself after writing a paragraph, "What can I remove from this paragraph without destroying its meaning and its relationship to what precedes and what will follow?" Then, ask another question, "Does my reader require all that data to comprehend, evaluate or make a decision with?" If you find you can do without excess words, excess description and excess supportive data, you will end up with a tighter, better and more informative report.

These principles should also be used to evaluate graphs, photographs, diagrams and other illustrations. Remember, illustrations should support or aid comprehension rather than being a crutch on which your report leans. The same should be kept in mind when determining just how much you should append to your report. There is no need to copy all your lab notes to show that detailed experimentation was performed to substantiate the results. A statement that the notes exist and are available will suffice.

Pitfall 9: Not highlighting

Again, I believe the analogy of the painter applies. A good painter also knows what to highlight and what to subdue in a portrait or scene.

If you don't accent the significant elements, findings, illustrations, data, tests, facts, trends, procedures, precedents, or experiments pertinent to the subject and object of your report, you place the burden of doing so on your reader. As a result, he may consider the report a failure, draw his own conclusions, or hit upon the significant elements by chance. In any event, don't leave it up to your reader to search out the major points of your report.

Highlighting is one step past knowing what goes into your report and what to leave out (see Pitfall 8 above). All the key points of your report should define and focus on the purpose of your report. They must be included in your summary or conclusions, but these sections are not the only places to highlight. Attention should be called to key elements needed for the understanding of your material throughout the body of the report. Several methods may be used: you can underline an important statement or conclusion, you can simply point out that a particular illustration is the proof of the results of an experiment, or, as most professional writers do, you can make the key sentence the first or last sentence of a paragraph.

Pitfall 10: Not rewriting

Did you ever hear of an actor who hadn't rehearsed his lines before stepping before an audience? An actor wouldn't chance it—his reputation and his next role depend on his performance. The engineer shouldn't chance it either. Don't expect the draft of your report to be ready for final typing and reproduction without rewriting.

Once you have judged what your report will contain and how it will be organized, just charge ahead and write the first draft. Don't worry about choosing the precise word, turning that meaningful phrase, or covering all the facts in one paragraph or section. Once you have written your first draft (and the quicker you accomplish this the more time you will have to perfect the text), you are in a better position to analyze, tailor, and refine the report as a whole. Now you are also able to focus all the elements toward your purpose and your audience.

As you begin the rewriting process simply pick up each page of your draft, scan it, and ask yourself what role the material on that page plays in the fulfillment of the report's objective and understanding. You will find that this will enable you to delete, add, change and rearrange your material very quickly.

After you have completed this process, then rewrite paragraph by paragraph, sentence by sentence, and word by word. Your final step is to repeat the procedure of examining each page's contents. When you are satisfied with its flow and cohesion, then you will have a good report, one you know will be well received and acted upon.

5

Resumes and letters of application

Communicating your message through letters and resumes

JOHN L. MUNSCHAUER

John L. Munschauer is Director of Career Development Services at Cornell University and author of numerous articles on the benefits of the liberal arts to career planning.

Why use a resume?

The purpose of a resume is to convey a message, a purpose easily forgotten in the ritual of preparing it. At every turn, you will get conflicting advice about how to conduct the ritual:

You must have a resume to get a job.

The purpose of a resume is to get an interview.

Every resume must have a job objective.

A resume should never, ever be longer than one page.

On a resume, list experience chronologically.

On the other hand, others advise:

Don't use a resume if you are looking for an executive position or merely seeking information.

Resumes typecast you and narrow your options.

Interview for information first, determine what an employer wants, and then decide whether offering a resume will suit your purposes.

Employers want to know what you have done; list your experience by function, not chronologically.

Two or more pages present no problem if your resume follows a logical, easy-to-read outline.

Reprinted with permission from *Jobs for English Majors and Other Smart People,* © 1982 by Peterson's Guide, Inc., P.O. Box 2123, Princeton, NJ 08540. Available in bookstores or direct from the publisher.

The more opinions you get, the more confused you become, but you finally work up something. You send it out. You get little or no response. You change your resume from one page to two pages—or from two pages to one. The results are no better. Somebody says to try pink paper, *that* will get attention. You decide you should have used blue. You fiddle and fiddle with your resume, trying to find the magic formula that will get you what you want. You are caught up in the ritual, forgetting that the purpose of a resume is to send a message. It is hard not to be distracted from that purpose. You worry about style. You concentrate on developing a format and forget the message, instead of concentrating on the message so you can develop a format that will bring it out. In developing a resume, it is easy to get wrapped up in yourself, because a resume is a document about yourself. The result: a resume more suitable as an obituary than as a message.

What message? You can get so caught up in the ritual that you can even forget to ask fundamental questions. What do I have to say? What is the best way to say it? The best way is not always the obvious one. When you want an information interview, might you not give your message best in person or via the telephone? In such cases, the less you commit to paper, the more flexibility you will have in an interview to adjust the message to the circumstance.

In fact, there may be times when you should think in terms of visual messages rather than regular resumes. Take the case of Cheryl Fender, an alto who learned that the Springfield Opera Company was auditioning for an alto to sing *Carmen*. Cheryl was well qualified, and her voice teacher was well known as a coach of only the most gifted singers. On the other hand, her resume, which was incomplete, with gaps in her history, showed extensive experience as a secretary. Wisely, she did not send the resume, because it might have established her as a secretary who wanted to sing, rather than as a singer who had supported herself by being a secretary. Instead, she asked herself, What is important to a director? Voice, of course, so she outlined her training in a letter. But dramatic ability also mattered, so she included pictures of herself on stage in roles that beautifully illustrated her dramatic ability. She got an audition. Cheryl followed a good marketing rule: Don't confuse customers by flaunting things that don't speak to their needs.

Giving your message

Even in an interview, it is difficult to use the language of employment so that employers can hear what you are saying in terms of their needs; but in letters and resumes, you almost automatically say "I want," making it doubly difficult to send messages that speak to employers' needs. It won't do to write, "I have analyzed my qualifications and have determined that they fit your needs." In essence, that is what most letters of application say. It doesn't work. Yet you don't know how to use "I" words that come out saying "You need me." If you understood the language of employment, you would find it easy to do.

I recall being asked to review an applicant's proposed application letter for a job on the staff of a yachting magazine. The letter was beautifully written, but it left the impression of being all "I": "I want to write so very much, and I am sure I can learn if you give me a chance . . . I got straight A's in English . . . I love boats . . . I am a sailor . . . I was captain of the sailing team in college. . . ."

What do you say to a person who has written a letter like that? He had obviously spent hours composing it. In its appearance and use of language, it met the highest standards. I could not squelch the young man's hopes, so I had him read *Book Publishing,*[1] a pamphlet Daniel Melcher wrote when he was president of R. R. Bowker Company. Although the pamphlet is concerned only with book publishing, I thought its message would be useful for someone interested in magazines as well. Let me paraphrase a portion of Melcher's pamphlet:

> I like publishing and you might like it too. It is only fair to warn you, however, that publishing attracts a great many more people than the industry can possibly absorb. Sometimes it seems as though half of the English majors in the country besiege publishing offices for jobs each year. While we hope that you will have something to offer us, you might as well face it. Publishers are experts in the art of the gentle brush-off.
>
> The interviewer hopes that you have what he needs, but it turns out that you have never looked into any of the industry's trade journals nor read any books about the industry. You haven't even acquainted yourself with the work of your university press. You tell the interviewer that you are willing to start anywhere, but it develops that a file clerk's job would not interest you, you do not want to type, you don't think you can sell, and you know nothing about printing.
>
> The fact is, all you have thought about is what you want, but it is his needs that create jobs, and you must address yourself to needs.
>
> Your problem, therefore, is to learn as much as possible about the industry before you go looking for a job. Only in this way will you be able to put yourself in the publisher's place and talk to him about his needs rather than about your wants.

After reading Melcher's advice, the fellow said he got the point, thanked me, and left, returning about three weeks later to show me his revised letter. It began in much the same way as his earlier effort—with "I's": "I majored in English . . . I have done considerable sailing . . ."; but after a few short sentences, he suddenly changed his tack. A new three-sentence paragraph began like this: "With my interests, naturally I want to work for you. But more to the point is not what I need but what you need." He followed this with three

---

[1] Daniel Melcher, *Book Publishing* (New York: Publishers Weekly). Pamphlet, out of print.

words that many women refuse to use and that men almost never think to use: "I can type." Right away, he showed that he knew he must be useful.

So much for the easy part of the letter. Next came the difficult part. Although he had had no experience, and had never submitted an article to a magazine or even written for a student newspaper, he still had to come up with something that would interest the editor. He found the solution. The letter continued in this manner.

> . . . In looking into the field of journalism, I visited the editor of our alumni magazine, and I talked to magazine space salesmen and to executives responsible for placing ads in magazines. I also visited a printer who has contracts with magazines. In addition, I have been reading trade journals and several books on the industry. As I looked into publishing, it occurred to me that of all the things I have done, the one I could most closely relate to the field was, strangely enough, an experience I had as a baby-sitter.

Immediately, he had the editor's full attention. How could baby-sitting fit in with publishing? During the summer of his junior year in college he had taken a job as a sailing instructor, tutor, and companion for the children of a wealthy family that summered on the coast of Maine. The parents often went away for a week or more at a time, leaving a governess in charge, with a cook, chauffeur, maid, and gardener to do the chores and the student to keep the children busy. While the parents were on a cruise, the governess suffered a stroke, sending the cook into a tizzy, the maid into tears, and the chauffeur and gardener to the local bar. Only the student could cope, and he took charge and managed the establishment for the rest of the summer.

In his letter of application he described the crisis, and subsequent problems he had faced, and told how he had met them. Then he related those experiences to the problems that he had learned editors, advertisers, printers, and others encounter in the publishing industry. Reading this letter, you could picture the student working for a publisher. There would be no slipups with the printers. Advertisers and authors would be handled with tact, yet he would get them to turn in their copy on time. He came through as someone with ingenuity, energy, and reliability; and the letter itself testified to his writing ability.

Did he get the job? Yes, and no. He got an offer, but because of the publisher's urgent needs the job had to be filled immediately, and he could not accept it. He was teaching school at the time and felt that in fairness to his pupils he should finish the school year. However, a while later, the letter surfaced again when a group of editors at a meeting chatted during lunch. The subject of good editorial help came up. It followed the usual theme, "They don't make 'em like they used to." The editor of another sailing magazine complained that he had been looking for an assistant, but despite hundreds of applicants had found none suitable. At this point, the editor of the yachting magazine described the young man's letter and agreed to share it. The upshot was, again, a job offer. This time the timing was right, and the young man took the job. There is nothing quite like the staying power of a well-written letter. It is remembered.

Later, I complimented the young man, telling him I had never read a better letter. "It was easy," he said.

> The first letter was the tough one. I didn't have anything to say other than "I have a good record. Please give me a chance," but I knew everyone else was saying the same thing, so I would have to say it better. I struggled with every word, trying to make an ordinary message extraordinary, but even elegant words can't make something out of fluff. I didn't known anything about publishing, so I didn't have anything to say to publishers; nor could I be really convincing without knowing enough about the work to decide whether I wanted that kind of job or not. But publishing sounded exciting, so I thought I would give it a fling.

The importance of knowing what the job is all about

"When I looked into the field in depth," the young man continued,

> I became confident that I had something to offer. Thanks to a few good high school teachers and college professors, I knew where to place a comma and a colon, so I had a technical skill to offer publishers. And I knew sailing. But the big need I saw was one I discovered when I was teaching school and again when I was an assistant manager at a McDonald's—a need for people who can get things done. Such a quality is hard to describe without using a situation. I could have alluded to any number of jobs I had held, but I chose the baby-sitting job because I thought it would be different and would introduce an element of surprise. Apparently, the analogy worked. I got a job. More important, I wanted it. If I had received a job offer after sending the first letter, before I had really investigated publishing, I would have taken the job with an attitude of "teach me." That's a passive role, one of an observer, and observers tend to be critical. The chances are fifty-fifty that I would not have known how or where to contribute and would have quit after a while. Instead of offering my employer solutions to his problems, I would have become one of his problems.

I asked the young man if he had used any other supporting documents, such as a resume, to help him get the job. "I had a resume," he replied,

> but I held it back, because my task was to transfer the qualities I had demonstrated as a baby-sitter to the needs of a publisher, and I couldn't seem to do that in a resume. A resume is a good way to outline facts, but I had to use prose to develop the analogy.

It occurred to me that I might be asked for a writing sample, so I wrote a couple of articles about sailing, but again I held them back. Later, I was glad, because I wasn't asked for writing samples. Editors, I discovered, have very definite ideas about what they want. No matter how good my articles were, there was a good chance the editor wouldn't have liked them and would have rejected me along with my writing. I once sold pots and pans door-to-door and I had a customer ready to buy, but I hadn't yet given him my big sales argument, which had to do with waterless cooking and saving vitamins, so I proceeded to give him the good news. "That's the biggest line of nonsense I ever heard," he said, and practically threw me out the door. My sales manager told me I had learned a good lesson: quit when you are ahead. That's a sales principle I have adapted to job hunting.

Are these examples intended to be good arguments for not using resumes? No. They simply emphasize that it is important to determine the best way to get a message across. There are times when there is no substitute for a resume. When employers advertise and list the qualifications they seek, there is no better way to respond than to send a resume outlining qualifications.

Letters of application

Sometimes, however, it is hard to imagine how to make the letter of application you send with your resume more effective than those of the hundreds of others who are responding to the ad. Imagine being on the receiving end of applications at General Motors or Exxon, organizations that list jobs in the *College Placement Annual*,[2] which is given to almost every college graduate in the country.

To find out about the effectiveness of letters and resumes, I visited corporations and asked employment managers for their comments. "Here," said one employment manager, as he picked up an eighteen-inch stack of letters and handed it to me. "This is my morning's mail. Read these letters and you'll have your answer."

"I can't," I protested. "I have only two hours, and there's a day's reading here."

"Yes, you can," he replied. "Unfortunately, you'll get through the pack in a half hour, because a glance will tell you that most are not worth reading."

It was hard to believe that the letters could be that bad, but he was right. The typical letter was an insult. Among the letters that I did not finish reading was one on onionskin paper, in very light type—it must have been the fifth carbon in the typewriter. It began:

[2] *College Placement Annual* (Bethlehem, Pa.: College Placement Council, Inc.). Annual.

Dear Sir:

I am writing to the top companies in each industry and yours is certainly that. I want to turn my outstanding qualities of leadership and my can-do abilities to . . .

Enough of that. Also, the applicant hadn't even bothered to type in the employment manager's name, which he could easily have found in the *College Placement Annual.*

The next letter was written in pencil on notepaper. There may have been an Einstein behind that one, but I can't imagine anyone taking the time to find it out. Many other letters were smudged and messy. Some applicants tried to attract attention with stunts, such as putting cute cartoons on their letters. One piece of mail contained a walnut and a note that read, "Every business has a tough nut to crack. If you have a tough nut to crack and need someone to do it, crack this nut." Inside the nut, all wadded up, was a resume. Cute tricks and cleverness don't work at the General Mammoth Corporation.

At the same time, the good letters stood out like gold. Five letters, only five letters in that pile of hundreds, were worth reading. They had this in common:

• They looked like business letters. Their paragraphing, their neatness, and their crisp white 8½" x 11" stationery attracted attention like good-looking clothing and good grooming.

• They were succinct.

• There were no misspellings or grammatical errors.

As I read them, I heard a voice—the voice of a fusty old high school English teacher—commanding out of the past:

• If you can't spell a word, look it up in a dictionary.

• Use a typewriter. Pen and ink are for love letters.

• Clean your typewriter. Avoid fuzzy type. You wouldn't interview in a dirty shirt, so don't send a dirty resume or letter.

• For format, use a secretarial manual. If you don't have one, get one from the library. What do you think libraries are for?

How that English teacher would have loved the following advice from Malcolm Forbes, the editor-in-chief of *Forbes Magazine:*

Edit ruthlessly. Somebody ~~has~~ said that words are ~~a lot~~ like inflated money—the more ~~of them that~~ you use, the less each one ~~of them~~ is worth. ~~Right on~~. Go through your entire letter ~~just~~ as many times as it takes. ~~Search out and~~ Annihilate all unnecessary words, ~~and~~ sentences—even ~~entire~~ *paragraphs.*[3]

The following letter is typical of those I saw that day. Give it the Forbes treatment, and see what you can do with it. You may need a scissors as well as a pencil.

Dear Mr. Employer,

I am writing to you because I am going to be looking for employment after I graduate which will be from Michigan where I studied Chemical Engineering and I will be getting a Bachelor of Science degree in June. The field in which I am interested and hope to pursue is process design and that is why I am writing your company to see if you have openings like that. I think I have excellent qualifications and you will find them described in the resume which I have attached to this letter.

The five letters that stood out favorably were characterized by their simplicity. Here is a letter, fictitious of course, but enough like the letter I remember to give you an idea of the ones that created a favorable impression:

February 1, 1981

Mr. Paul Boynton
Manager of Employment
The United States Oil Company
1 Chicago Plaza
Chicago, Illinois 60607

Dear Mr. Boynton:

This June I will receive a Bachelor of Science in Chemical Engineering from the University of Michigan, and I hope to work in process design or instrumentation. I was glad to see your description in *Peterson's Annual Guide to Careers and Employment of Engineers, Computer Scientists, and Physical Scientists,* soliciting applicants with my interests. I am enclosing a resume to help you evaluate my qualifications.

While I find all aspects of refining interesting, my special interest in process design and instrumentation developed while working as a laboratory assistant for Professor Juliard Smith, who teaches process design. I wrote my senior thesis under him on the subject of instrumentation, and part of what I wrote will be used in a textbook he is writing and editing.

Would it be possible to have an interview with you in Chicago during the week of March 1? To be even more specific, could it be arranged for 10 A.M. on Tuesday, March 3? I am going to be in Chicago

---

[3] Malcolm Forbes, *How to Write a Business Letter* (Elmsford, NY.: International Paper Co., 1979). Handout and advertisement. [*Editor's note*: The whole of Forbes' advice is included in Part 3 of this anthology.]

that week, and this time and date would be best for me, but of course I would work out another time more convenient for you. In any event, I will call your office the week before to determine whether an interview can be worked out.

                                        Sincerely yours,

                                        Charles C. Marion
                                        1 Riverview Lane
                                        Ann Arbor, Michigan 48106

Marion's effective letter, and the four others, followed a pattern:

(1) The first paragraph stated who the writer was and what he wanted;

(2) The second paragraph, sometimes the third, and in one case a fourth paragraph, indicated why the writer was writing to the employer—mentioned areas of mutual interest, special talents that might be of interest to employers, or other factors relating to his qualifications that could be better described in a letter than in a resume; and

(3) A final paragraph suggested a course of action.

In the sample, Marion is very specific about what he wants. In the second paragraph of a letter, an English major seeking a job in marketing alluded to his experiences working in gas stations to show that he had learned something about marketing in the petroleum industry. He also told how he had started a new chapter for his fraternity, tying that experience into opening new gasoline stations. One of the other good letters, more of an inquiry than an application, concluded with, "I hope my qualifications will be of interest to you. If so, I look forward to hearing from you."

Hard work and professional typing make a good letter

While only the five letters were effective, the rest of those who wrote could have done as well. The point is they didn't. Most people who write letters in the future are not going to write good letters either. Therein lies your opportunity, because, like Marion, you can write letters that set you apart. You don't have to create a literary masterpiece; just don't hastily knock off a letter with thoughts that wander all over the page. Write it and rewrite it, following Forbes' advice. Unless you are an exceptional typist, you are not good enough to type it yourself. Hire a professional. Also, get an English teacher or someone in the word business to check your spelling, punctuation, and grammar. But be sure it is your letter. Somehow, a ghost-written letter always has a phony ring to it.

Don't delegate the job of letter writing

More important than style, however, is the thought process you must use in preparing letters and resumes. Don't short-change yourself by delegating your thinking to someone else. When you write to employers, you must think about their needs; then you must think about yourself and what you offer, and you should relate this to what you would like to do. The act of putting your thoughts on paper—thoughtfully—will make you sort out your ideas and interrelate them. When you see them on paper they will talk back to you, at times to suggest better ideas, at other times to tell you that you are off the mark. To organize your ideas, create an outline. In other words, prepare a resume even if you decide not to use it. The value of resume is frequently more in its preparation than in its use.

When you do give an employer your resume, make it a testimony to your ability to organize your thoughts. Remember, too, it must look sufficiently attractive to get an employer to read it. Unfortunately, most of the resumes I saw on employers' desks were just as unattractive as the letters; they had sloppy, crowded margins, were poorly organized, and were badly reproduced. At least 30 percent of the resumes had been put aside with hardly a glance because their physical appearance was awful. The rest got a twenty-second scan to see if they were worth studying.

Resume preparation

Following are two resumes that pass the appearance test with flying colors. Let's see how they fare during a twenty-second scan and beyond.

Nancy Jones

Nancy Jones's resume has arrived at the desk of a laboratory director who needs an assistant to help run a quality-control laboratory in a pharmaceutical company. With candidates far outnumbering openings in biology, the advertisement for the job has brought in hundreds of applications, and the director is wearily scanning them one by one to find the few that will be of interest to him. Conditions are not favorable for Nancy. She has to catch his eye with impressive qualifications or she is not going to get anywhere.

The director picks up Nancy's resume. Immediately, he is impressed, because it looks attractive. He thinks the resume reflects an orderly mind. Most resumes he has looked at just do not put it all together.

He begins to read. The job objective annoys him; it strikes him as being long-winded. Why couldn't she simply say she is interested in applied biology?

NANCY O. JONES

Present Address: After June 1, 1981:
105 Belleville Place   1212 Centerline Road
Ames, Iowa 50011   Old Westbury, New York 11568
Phone: 515-924-6674   Phone: 516-544-7119

Born January 6, 1959 5'7" 135 lbs. Single Excellent health

Career   Research and development in most areas of
Objective: applied biology, with an opportunity to work
                    with people as well

Education: Iowa State University, Ames, Iowa
                    Bachelor of Science, June 1981
                    Major: Biology   Concentration: Physiology

                    GPA: 3.3 on a 4.0 scale

                    Major Subjects   Minor Subjects
                    Mammalian Physiology   Qualitative Analysis
                    Vertebrate Anatomy   Quantitative Analysis
                    Histology   Organic Chemistry
                    Genetics   Biochemistry

Scholarships   University Scholarship: $2850/year
and Honors: Iowa State Science and Research Award

                    Dean's List two semesters

Activities: Volunteer Probation Officer, 1978-79
                    Probation Department, Ames, Iowa

                    Tutor, Chemistry and Math, 1978-81
                    Central High School, Ames, Iowa

                    Member, Kappa Zeta social sorority

                    Women's Intercollegiate Hockey Team

Special Skills: Familiarity with Spanish; PLC and FORTRAN
                    computer languages; typing

Work   Teaching Assistant and Laboratory Instructor
Experience: Freshman Biology School year, 1980-81

                    Waitress, Four Seasons Restaurant
                    Catalina Island Summers, 1979, 1980

The resume of Nancy O. Jones does not communicate her career-related experience.

What is this business about working with people? Is that there because she has doubts about biology?

If resumes are supposed to say only what needs to be said, what about this line?

```
Born January 6, 1959  5'7"  135 lbs.  Single  Excellent health
```

Does it say anything about her ability to do the job? The biologist doesn't think so. His eyes move down the page:

```
Education:    Iowa State University, Ames, Iowa
              Bachelor of Science, June 1981
              Major: Biology  Concentration: Physiology

              GPA: 3.3 on a 4.0 scale

              Major Subjects        Minor Subjects
              Mammalian Physiology  Qualitative Analysis
              Vertebrate Anatomy    Quantitative Analysis
              Histology             Organic Chemistry
              Genetics              Biochemistry
```

She has used the outline form well, so he is able to take in a great deal of information in one look. She impresses him with her education.

Double spacing above and below her grade point average makes it stand out. However, if her average had not been quite as good, and if she had not wanted to feature it, she could have used single spacing to make it less conspicuous, like this:

```
Education:    Iowa State University, Ames, Iowa
              Bachelor of Science, June 1981
              Major: Biology  Concentration: Physiology
              GPA: 3.3 on a 4.0 scale
```

Now that she has told him about her grades and her courses she wants to drive home the point that she was no ordinary student. She flags him down with the headline "Scholarships and Honors," and he sees her financial aid and science awards, which make a favorable impression. Two times on the dean's list may not be important enough to set apart by double spacing, but it makes a modest impression on the biologist.

Up until now, the biologist's overall impression is favorable. His eye is ready to take in the next batch of information:

```
Activities:        Volunteer Probation Officer, 1978-79
                   Probation Department, Ames, Iowa

                   Tutor, Chemistry and Math, 1978-81
                   Central High School, Ames, Iowa

                   Member, Kappa Zeta social sorority

                   Women's Intercollegiate Hockey Team
```

She almost loses him by featuring her work as a probation officer. That is not of primary interest to him, but through good spacing and placement, she succeeds in drawing his eye to the next item, which states that she has tutored chemistry and math. Unfortunately, Nancy now loses him permanently by ranking tutoring along with the sorority and the hockey team. He guesses she has made her major statement about biology, and so he turns to the next resume.

Good as it is, Nancy Jones's resume could be improved by reorganization. Her tutoring in chemistry and her two years as a teaching assistant and laboratory instructor in biology reveal more than an academic interest in biology. They should be featured. Everything related to biology and any other information of possible use to the employer should be put under a new marginal headline and statement, as follows:

```
Career-related    Biology Lab Instructor and Teaching Assistant
Experience:       Freshman Biology (1980-81)

                  Chemistry and Mathematics Tutor
                  Central High School, Ames, Iowa (1978-81)

Skills and        Microscopy              Computer Languages:
Interests:        Electron Microscopy       FORTRAN, PL/1,
                  Histology                 COBOL
                  Spectrum Analysis       Statistics
                  Small-Animal Surgery
```

Now, the director is able to see things she can do. "Good," he says to himself, "She can use an electron microscope. We need someone with that skill."

By no means should Nancy eliminate mention of the hockey team, but since she is applying for a professional job, the first bait to throw is credentials; they testify to her ability to do the work. After that, what may sink the hook is how the employer sees her as a person. He may have picked out bits here and there that testify to her diligence, but the picking out depends on chance reading; it would be best not to leave anything to chance. With a slightly different presentation, she might ensure that he receives an impression of her diligence.

There are other areas she could strengthen, things she has underplayed or totally neglected to mention. Sticking the waitress job in down at the bottom of the page is almost an apology for it. She may also have had other jobs, such as baby-sitting, household work, or door-to-door sales, that she belittles in her mind and has not even mentioned. Mention of such things might make a good impression on an employer looking for somebody who is not afraid to work and who is mature for her years.

If we quizzed Nancy, we might find that she could put something like this on her resume:

```
Scholarships      90% self-supporting through college as follows:
and Financial     University Scholarship: $2850/year
Support:          Iowa State Science and Research Award


Work              Teaching Assistant and Laboratory Instructor
Experience:       Freshman Biology  School year, 1980-81

                  Waitress, Four Seasons Restaurant
                  Catalina Island  Summers, 1979, 1980
```

And this, because her activities tell something about her as a person:

```
Activities:       Volunteer Probation Officer (1978-79)
                  Kappa Zeta social sorority
                  Women's Intercollegiate Hockey Team
                  Skiing, sailing, singing, tennis
```

Double spacing has been cut down so as not to overemphasize the less important items, yet a string of other things not terribly important in themselves have been inserted to support the impression of an active, interesting person. Sometimes you want to leave an impression, at other times you want to emphasize a qualification. For example, Nancy wanted to feature her studies, and the way she brought them out by listing them in a column was good. If she had listed them like this, they would not have stood out:

```
          Major Subjects: Mammalian Physiology,
          Vertebrate Anatomy, Histology, Genetics
          Minor Subjects: Quantitative Analysis,
          Qualitative Analysis, Organic Chemistry,
          Biochemistry
```

On the next page, you will see the Nancy Jones resume as she might have revised it to emphasize the strong points that would have been of interest to that biologist looking for a resourceful assistant. The added emphasis might have made the difference that would have landed the job for her.

Nancy's resume is pure outline, devoid of prose. It works well for her. When she states that she has studied quantitative and qualitative analysis, and

NANCY O. JONES

Present Address:
105 Belleville Place
Ames, Iowa 50011
Phone: 515-924-6674

After June 1, 1981:
1212 Centerline Road
Old Westbury, New York 11568
Phone: 516-544-7119

Career Objective: Research and development in applied biology

Education:
Iowa State University, Ames, Iowa
Bachelor of Science, June 1981
Major: Biology   Concentration: Physiology

GPA: 3.3 on a 4.0 scale

Major Subjects	Minor Subjects
Mammalian Physiology	Qualitative Analysis
Vertebrate Anatomy	Quantitative Analysis
Histology	Organic Chemistry
Genetics	Biochemistry

Career-related Experience:

Biology Lab Instructor and Teaching Assistant
Freshman Biology (1980-81)

Chemistry and Mathematics Tutor
Central High School, Ames, Iowa (1978-81)

Skills and Interests:

Microscopy	Computer Languages:
Electron Microscopy	FORTRAN, PL/1,
Histology	COBOL
Spectrum Analysis	Statistics
Small-Animal Surgery	

Scholarships and Financial Support:

90% self-supporting through college as follows:

University Scholarship: $2850/year
Iowa State Science and Research Award

Waitress, Four Seasons Restaurant
Catalina Island (Summers, 1979, 1980)

Teaching and instructing, baby-sitting,
home maintenance, selling

Activities:
Volunteer Probation Officer (1978-79)
Kappa Zeta social sorority
Women's Intercollegiate Hockey Team
Skiing, sailing, singing, tennis

The revised resume for Nancy O. Jones effectively features her career-related experience in a separate section.

knows PL/1 and FORTRAN, a scientist reading her resume knows what this means.

Janet Smith

Janet Smith, whose resume follows, has a different problem in presenting her qualifications. She needs to *describe* what she did in order to tell an employer about her qualifications, so her resume calls for a mixture of key words and prose to get her message across. Yet prose can destroy the effect of an outline. The solution lies in imitating newspaper editors, who use headlines and sub-headlines to attract readers. Like a newspaper, a resume should lend itself to skimming, so the reader can quickly pick up a good overview of what is important. Then the reader can select specific things that interest him, and read further. There is an art to using headlines, but Janet Smith hasn't mastered it, at least not in the resume she used to apply for a job with Hermann Lang-felder, a hard-bitten old hand with thirty years in personnel and labor relations in the machinery business. He picked up her resume and read:

```
CAREER         A challenging position in personnel admin-
OBJECTIVE:     istration requiring organizational ability
               and an understanding of how people function in
               business and industry.
```

That was pure baloney, and he choked on it. He thought of the hours he had spent in meetings, listening to a lot of hot air. "Challenge, my foot!" he muttered. Then he read the bit about her organizational ability and her under-standing of how people function in business and industry. "I've been at this business for thirty years," he groused to himself, "and I still can't figure out how people function in industry. But *she* knows all about it."

Beware of misleading headlines

"Well, let's see what she's done," he said to himself, and then his eyes fell on "UNIVERSAL METHODIST CHURCH." That did it! He didn't want to bring any do-gooder into his factory to preach. He rejected her. And the fault was hers, in using the headline.

Janet's job at the church was administrative, not ministerial, and the church didn't care whether she was Jewish, Catholic, or agnostic. Only in its ministerial work does the Methodist Church need Methodists. But people—Langfelder and the rest of us—respond to symbols and make snap judgments on the basis of symbols. Janet had put the symbol of the church—its name—in a heading, when she could have done something better.

When you lay out your resume, think of symbols. Imagine you are a newspaper editor who wants to put a story across. As an editor planning headlines, you must imagine yourself in the position of the reader. Ask your-self, "What words will catch the reader's eye? What words will put him off?"

JANET V. SMITH
111 Main Street
North Hero, Vermont 05073
(802) 772-4325

CAREER
OBJECTIVE:
A challenging position in personnel administration requiring organizational ability
and an understanding of how people function in
business and industry.

EDUCATION:
PURDUE UNIVERSITY
Hammond, Indiana
Master of Industrial Relations June 1981

SMITH COLLEGE
Northampton, Massachusetts
Bachelor of Arts, Magna Cum Laude June 1976

WORK
EXPERIENCE:
UNIVERSAL METHODIST CHURCH
1 Central Square
New York, New York 10027
Assistant Personnel Officer. 1978-80
Responsible for interviewing applicants for
clerical positions within the organization and
for placing those who demonstrated appropriate
skills, for accepting and dealing with employees' grievances, and for developing programs on career advancement.

CORTEN STEEL COMPANY
10 Lake Street
Akron, Ohio 44309
Assistant, Personnel Office. 1976-78
Responsible for all correspondence of Personnel Director and for interviewing some
custodial applicants and referring them
to appropriate supervisors for further
interviewing.

BORG-WARNER, INC.
Ithaca, New York 14850
Assembly linewoman. Summer 1975
Assembled parts of specialized drive chains in
company with thirty other men and women.

COMMUNITY
SERVICE:
PLANNED PARENTHOOD
North Hero, Vermont
Counselor. Summers of 1973 and 1974
Explained various aspects of family planning
and provided birth control information to cli-
ents of Planned Parenthood. Made referrals to
other counselors and physicians where
appropriate.

AUXILIARY
SKILLS:
French: Fluent. Knowledge of office procedures.
Knowledge of IBM 360 Series

The resume of Janet V. Smith communicates irrelevant messages.

Use words that fit the job in question, and play down those that can lead an employer to think of you in terms that don't relate to the job. Ask yourself, "Does this say something to the employer?" Janet Smith missed the obvious. Her job titles made perfect headlines to attract Langfelder, and some of her duties yielded key words that sent appropriate messages. The following arrangement would have been more effective for her:

```
WORK            ASSISTANT PERSONNEL OFFICER
EXPERIENCE:     1978-80
                Universal Methodist Church
                1 Central Square
                New York, New York 10027
                Interviewing, placement, grievances, and
                training of applicants and employees in the
                clerical and support services of the church or-
                ganization. Developed programs for the career
                advancement of employees.

                ASSISTANT, PERSONNEL OFFICE
                1976-78
                Corten Steel Company
                10 Lake Street
                Akron, Ohio 44309
                Interviewing, referring, and correspondence
                as an assistant for Personnel Director. Re-
                sponsible for all correspondence and for in-
                terviewing some custodial applicants and
                referring them to supervisors for further
                interviews.
```

Let's suppose that Janet Smith has a chance to submit a second resume to Langfelder. This time, she uses headlines that pinpoint the ideas she most wants to get across, so that he makes it past the Universal Methodist Church and gets down to the assembly line (see p. 252). "Hey, now, look at that!" he thinks. "She's worked out there on the floor. That means she's heard all the language and knows the gripes and the tedium. We have a lot of women in this factory, and it might be a good thing to have a down-to-earth, smart woman on my staff." (His "sexism" may have been showing, but this employer is alive and kicking somewhere out there, and you may have to deal with him.)

Mark Meyers—the functional resume

Janet Smith and Nancy Jones were lucky. Their training and experience translated into satisfactory headlines to highlight their experience, but that doesn't always work. Sometimes the only way to get a message across is to create a resume based on functions. Mark Meyers, whose resume is on page 253, adopted that technique to help him get a job in community recreation.

When he began to write a resume, he tried time and time again to get his message across in a conventional resume in which he first listed his education, then his experience in chronological order, and finally his activities, hobbies,

JANET V. SMITH
111 Main Street
North Hero, Vermont 05073
(802) 772-4325

CAREER
INTERESTS: Personnel Administration and Labor Relations

EDUCATION: PURDUE UNIVERSITY June 1981
                  Hammond, Indiana
                  Master of Industrial Relations

                  SMITH COLLEGE June 1976
                  Northampton, Massachusetts
                  Bachelor of Arts, Magna Cum Laude

WORK   ASSISTANT PERSONNEL OFFICER. 1978-80
EXPERIENCE: Universal Methodist Church
                  1 Central Square
                  New York, New York 10027
                  Interviewing, placement, grievances, and
                  training of applicants and employees in the
                  clerical and support services of the church or-
                  ganization. Developed programs for the career
                  advancement of employees.

                  ASSISTANT, PERSONNEL OFFICE. 1976-78
                  Corten Steel Company
                  10 Lake Street
                  Akron, Ohio 44309
                  Interviewing, referring, and correspondence
                  as an assistant for Personnel Director. Re-
                  sponsible for all correspondence and for in-
                  terviewing some custodial applicants and
                  referring them to supervisors for further
                  interviews.

                  ASSEMBLY LINEWOMAN. Summer 1975
                  Borg-Warner, Inc.
                  Ithaca, New York 14850
                  Factory work experience on an assembly line.
                  Worked with a team of thirty other men and
                  women.

COMMUNITY   COUNSELOR. Summers 1973 and 1974
SERVICE: Planned Parenthood
                  North Hero, Vermont 05073
                  Explained various aspects of family planning
                  and provided birth control information to
                  clients of Planned Parenthood. Made referrals
                  to other counselors and physicians where
                  appropriate.

AUXILIARY   French: Fluent. Knowledge of office procedures.
SKILLS: Knowledge of IBM 360 Series

The revised resume for Janet V. Smith stresses what she did rather than the less important point of where she did it.

MARK MEYERS

1414 South Harp Road
Dover, Delaware 19901

CAREER INTEREST:
  Community recreation.

RECREATION PROGRAMMING EXPERIENCE:
  Planned and implemented programs in stagecraft and drama;
    assisted with programming in ceramics, photography, and
    physical fitness for Dover Youth Bureau summer program.
    Summer, 1976.
  Lectured and led tours at Atlantic County (Delaware) Park
    nature trail and visitors' center. Prepared slides to
    illustrate lecture; helped in construction of nature
    exhibits. Summer, 1975.
  Coordinated men's intramural sports competitions for Ho-
    boken University. Had responsibility for equipment and
    scheduling. 1974-76.

PUBLIC RELATIONS AND PROMOTION EXPERIENCE:
  Directed publicity efforts of University Drama Club for
    several productions. Developed innovative techniques —
    such as a costumed cast parade to arouse interest in an
    avant-garde staging of ''Alice in Wonderland,'' which
    gained campuswide attention. 1974-75.
  Advertised in various media and became familiar with ad-
    vertising methods, including writing news releases, tap-
    ing radio announcements, designing graphics for posters
    and fliers. Drama Club, 1975-77; men's sports, 1974-77.

LEADERSHIP AND ATHLETIC ABILITIES:
  Trained in outdoor leadership and survival skills by
    Outward Bound. Summer, 1973.
  Coached hockey and basketball. Dover Youth Bureau, 1976.
  Participates in hockey, swimming, backpacking.

RESEARCH AND EVALUATION ABILITIES:
  Prepared college research project on the recreational
    needs of residents of a Hoboken neighborhood. Designed
    questionnaire to solicit residents' own perceptions of
    their needs; interviewed residents and local officials.
    Fall semester, 1976.
  Reported on effectiveness of Dover Youth Bureau program-
    ming. Summer, 1976.

OFFICE SKILLS:
  Dealt with unhappy and irate customers at McGary's Depart-
    ment Store, Dover. Worked at service desk, tracking down
    problems and rectifying errors. Summer, 1972.
  Typing and use of office machinery.

EDUCATION:
  Bachelor of Science, June, 1977   Major: Recreation
  Hoboken University, Hoboken, New Jersey Minor: Drama

References are available upon request.

The resume of Mark Meyers illustrates the functional style.

and interests. But writing it conventionally raised all sorts of problems. He wanted to highlight his public relations and promotion experience, some of which he had been paid for and some not. Some of it had also been secondary to a primary assignment. Dividing up this experience and placing bits and pieces of it in various parts of the resume to make it conform to a conventional style diluted its impact. Also, his athletic ability and experience would mean a great deal to an employer in his field, but how could he show it effectively? Some of it had been gained as a participant, some through training, and some as a coach. Could he expect an employer to sift through the various sections of the resume to find out all he had done in athletics? (Remember, you can only count on an employer giving a resume a twenty-second scan before deciding whether or not to study it more fully.)

His solution, as you see, was to outline the functions of the job he wanted and then complete the outline by listing things he had done that pertained to each area. Thus, under each function he developed the equivalent of a mini resume.

Preparing a resume for a specific job

Mark stated his case well, but you can't get blood from a stone. He had to look outside his field, because jobs in it were virtually nonexistent due to cutbacks in government funding. In his search, he ran across the following job listing from the publisher of a magazine for parents:

EDITORIAL SECRETARY

B.A. in Liberal Arts

Interested in childhood training. Well organized, outstanding language skills. Typing and clerical skills,
potential to use electronic text-editing equipment.
Reporting to Coordinating Editor, _____ _____
magazine. Assist in all editorial functions. Evidence
of creativity essential. Entry-level position with
career potential.

Can you put yourself in his shoes, analyze the job, and devise a resume that speaks to the stated needs of this employer? The clue to doing it is to go through the job description and step by step take your cue from the employer. Right off you will hit a bit of a snare because the employer has specified a B.A., while Mark has a B.S. His drama minor might give him appropriate credentials, however. If he shows his education something like the example below, it might reflect the liberal background the employer apparently prefers:

```
EDUCATION:   HOBOKEN UNIVERSITY, B.S., 1977
             Major: Recreation    Minor: Drama

             Humanities courses:

             Introduction to Dramatic Literature
             British Drama to 1700
             History of Theater
             Playwriting
             Introduction to Poetry
             Shakespeare
             English History
```

Next, the job calls for an interest in childhood training, then language skills, and so on. Each specification suggests a headline for a resume. Mark faces another stumbling block in being able to demonstrate an interest in childhood training, since his experience was only with older youths. Since a resume makes points by stating facts, he cannot demonstrate an interest in childhood training in his resume because he lacks the appropriate experience. However, he *can* describe his interest in the letter that usually goes hand in glove with a resume. He has a good basis for doing it, because the field of recreation certainly has much to do with the entire range of human development from childhood on. He should be able to point out corollaries in his education and experience with the work being done in childhood training, and he could check a library for information about childhood training to help develop the corollaries. Above all, he should read the magazine and try to tie in as many of his experiences as possible with the purpose of the magazine.

With his interest in childhood training brought out in a letter to complement his resume, he might then proceed to develop his outline as follows:

```
EXPERIENCE:   HUMAN DEVELOPMENT

              Dover Youth Bureau. Planned and imple-
              mented programs in drama, photography,
              athletics, health. Summer, 1976.
              Research Project. Studied recreational
              needs of a Hoboken neighborhood. Inter-
              viewed residents, developed question-
              naire. Project provided an insight into
              the family life and problems of parents
              in a neighborhood setting. Fall semes-
              ter, 1976.
              Outdoor Leadership Training. Practical
              experience in human development
              through a knowledge of nature and sur-
              vival skills.
```

LANGUAGE SKILLS AND CREATIVITY

Writing. Wrote releases, developed advertising, and prepared radio an-
nouncements for Drama Club, men's
sports programs, and other events.
Wrote report on Hoboken research project. 1974-77.
Lecturing. Gave talks and led tours at
Atlantic County (Delaware) Park nature
trail and visitors' center. Summer, 1975.
Audiovisual. Prepared slides for audiovisual presentation for visitors'
center. 1975.
Promotional. Created innovative tech-
niques--such as costume parade of cast
--to arouse interest in avant-garde
staging of ''Alice in Wonderland,''
which gained campuswide attention.
1975.
Design. Designed posters, fliers, and
other graphics for sporting events,
plays, and other campus events.
1974-77.
Constructed nature exhibits at nature
center. 1975.

CLERICAL AND ADMINISTRATIVE

Clerical. Typing and use of office machinery, McGary's Department Store.
Summer, 1972.
Administration. Responsible for equipment, scheduling of programs, and coordinating of competitions. Hoboken
University, 1974-76.

The functional resume allows you to develop a different message for each job, or type of job, you wish to apply for. Different functions can be high-lighted, depending on what the job requires, and your specific experiences rearranged under different headings. It gives you the flexibility you need if your experience has been varied and diverse.

Bruce Gregory Robertson—a resume reflecting an active mind and body

Finally, there are merit employers who are interested in candidates not so much for what they know as for what they can learn. They want to see a

BRUCE GREGORY ROBERTSON

Home Address   College Address
105 Comstock Drive   5 Erasmus Drive
Pierre, South Dakota 57501   St. Paul, Minnesota 55101
(605) 257-7236   (612) 020-2020

CAREER INTERESTS - I am interested in the marketing of products
                     or services in industries such as banking,
                     publishing, and retailing.

EDUCATIONAL   - Macalester College, St. Paul, Minnesota A.B.,
BACKGROUND   May 1982
                     Major: English   Minor: Economics
                     Honors: Dean's List   1981-1982

EMPLOYMENT   - Sales - Southwestern Publishing Company,
                     Memphis, Tennessee.
                     Educational dictionary sales to families
                     with high school pupils: summer 1980 in
                     Tucumcari, New Mexico; 1981 in Asheville,
                     North Carolina; assisted in training new
                     salesmen.

                     Administrative and Clerical - Temporary Help,
                     Inc. Typing and other business machine opera-
                     tions, bookkeeping, clerking, complaint ad-
                     justments; short-term assignments with auto
                     dealers, banks, real estate operations,
                     schools, and similar employers: summer 1979;
                     part-time 1980.

                  - Miscellaneous - Camp counselor, stock clerk
                     in grocery store, baby-sitter, newspaper boy.
                     From high school on have earned money for
                     clothes, travel, and purchase and mainte-
                     nance of an automobile.

ACTIVITIES   - Vice president, Seven-come-eleven investment
                     club
                  - Varsity basketball
                  - Coach and tutor, St. Paul Concordia Boys Club
                  - Debate Club
                  - Book and Bottle literary club

SKILLS   - Typing
                  - PLC and FORTRAN computer languages
                  - German: fluent

HONORS   - Dakota Interstate Scholarship
                  - Hubert H. Humphrey First Prize, Minnesota
                     Intercollegiate Debate

INTERESTS   - Public speaking and debate
                  - Investments
                  - Writing
                  - Parachute jumping

A resume designed more to reflect an active, energetic personality than specific experience.

resume that reflects an active person with an active mind. Bruce Gregory Robertson is such a person, and he has designed his resume as much to reflect his personality as his experience. See his resume.

Additional advice about resumes

No matter how you develop your message, test it before you send it to employers. Get friends to give you a critique of your resume, especially if they are in an occupation in which you hope to find a job. A word of caution, however. Unless you guide them, their critique will relate more to the ritual than to the message. One job hunter had modeled his resume after Janet Smith's. The critic took a red pencil and indicated that all the dates of employment should be placed in the lefthand margin. That would have been a good idea if it had indicated long years of experience with a company such as IBM, signifying that the man had had considerable experience with one of the best-managed companies in the country. But the dates in question referred to short-time summer jobs, which were of no consequence to the message and cluttered the margin with information that distracted from the headlines.

One way to get a resume criticized is to hold it up a few feet from the reader and ask for comments on its appearance. Does it look neat? Is the layout pleasing? Does it look easy to read? Is the print good-looking?

Next, give the critics the resume to read. Let them make all the comments they want. You may pick up valuable ideas for improving its style and layout, but be careful you don't get caught up in inconsequentials. What you really want is to have your critics look at the resume as if they didn't know you. You might even show them a resume with an alias, then ask:

- What qualifications does this person have?
- What do you see this person doing with these qualifications?
- What kind of an employer would want to hire this person?
- Does the resume project an image of a certain kind of person? What kind? Aggressive? Thoughtful? Energetic? What?

In other words, ask your critics the most important question about your resume: "What message do you get about me?"

The style and appearance of the cover letter

MARCIA R. FOX

Marcia R. Fox directs career counseling and job placement at New York University's Graduate School of Public Administration.

*T*o an employer, your cover letter is the sole clue to your proficiency with the English language. Not only must your letters be grammatically and typographically *perfect* but you must also write them in an appropriate style.

1. Aim for simple but precise language. For example: "I am a forthcoming master's in electrical engineering interested in an engineering position with your company." Such virtues of simplicity were ignored in another version of this applicant's cover letter.

> Having graduated from the University of Colorado with a master's in electrical engineering, I have decided to pursue a career in a company such as yours.

2. If there is a reason for your particular interest in a certain employer, say so directly and succinctly. For example: "Professor John Sears, my research adviser, has repeatedly told me of your leadership in the field of prenatal nutrition."

3. Strive for precise, specific language. Avoid trite clichés and empty phrases:

> Employment at a highly diversified company such as yours represents a *challenging and stimulating beginning* to a career in business as well as the opportunity to significantly augment one's education.

> Although I am young, I have learned a great deal about human nature. I am in search of a *meaningful growth position* where I can use my talents and skills in a *responsible, challenging way*. I am a *creative thinker*.

Avoid vague terms such as "meaningful growth opportunity" and "creative thinker." Substitute "management trainee" or another specific term for "meaningful growth opportunity" or a "responsible" or "challenging" position. It's not the employer's job to figure out what you might find challenging; he has specific jobs to fill. Tell him what you want and assess whether or not the opportunity is meaningful at the interview!

4. Be concise. The best antidote for excessive wordiness is to read over a draft of your letter with a red pen in hand. Be prepared to justify each word. Notice, for example, how only the last sentence of the following paragraph is pertinent or substantive:

> I understand that your personal obligations in combination with the huge volume of applications through which you must wade, leave you with a limited amount of time to spend on each. With this in mind, thank you for your time and consideration.

In trying to identify with the employer's burden, the applicant unwittingly took up even more of the employer's time!

5. Use the active voice. The only way to sound energetic is to use the active voice. For example, state:

> "I would welcome the opportunity for an interview," not "An interview at your convenience is requested." Write, "I acquired a skill in economic analysis from these courses," not "From these courses, a skill in economic analysis was acquired."

An applicant's discomfort at actually writing a cover letter or even applying for a job can often result in excessive abruptness.

To help you hear the tone of your letter, wait a day before mailing it out, or have someone else read your letters aloud or silently.

Many professionals equate a professional style with a cool tone. The ideal cover letter, however, conveys the person behind the applicant. To create the rapport that will win you interviews, you must reveal some human warmth. Read the following example aloud to hear the tone of commanding coldness to avoid:

> Contact me if you have a suitable opening.

Do not be afraid to sound enthusiastic for fear it will be misinterpreted as feigned interest by a cynical employer. Always convey your enthusiasm professionally, by anchoring it to a credential or relevant fact. For example: "As a result of my fascination with gerontology, I obtained an H.E.W. fellowship to write a dissertation on the problems of indigent elderly women."

Remember that the employer is the stranger who must be persuaded to interview you. Problems of inappropriate tone may occur if you confuse him with another group or person. For example, naïve job hunters often mistakenly perceive the employer as a career counselor or as a "buddy" with whom to share fantasies. Don't close a letter this way: "I would very much appreciate talking with you so as to obtain a clearer perspective on my career goals." It is not the employer's responsibility to counsel you. Get the advice elsewhere!

A second problem arises when applicants approach the employer in a narcissistic way. To win interviews you must convince the employer you can help him rather than vice versa, as in this unfortunate example:

Employment at a highly diversified company such as yours offers many attractive health and vacation benefits. Moreover, it gives a challenging start to one who hopes to open his own business some day.

Some applicants mistakenly regard the employer as an audience in search of entertainment. This can lead to the kind of self-conscious "cuteness" exhibited in the following example:

Is there something special about a person who graduated six months early from both high school and college and has never cut a class in her entire educational career?

If you are tempted to point to any perceived inadequacies, remind yourself that employers are not confessors. It is self-destructive to point out facts that can harm you and which might otherwise have gone unnoticed. For example:

I am a forthcoming graduate of M.I.T.'s program in electrical engineering. While I did not win any significant research prizes, I . . .

Here is how a severe case of sour grapes was transmitted in a cover letter that manages to sound both confessional and hostile at the same time:

Academically, I stand well within the top half of my class. A failure to be glib on exams prevented me from qualifying for Phi Beta Kappa. However, this failure has not prevented me from succeeding in any other endeavor.

Sometimes it will be necessary to carefully stress the positive aspects of a situation that could easily have negative connotations to an employer. One woman reentered the job market using this bad opening sentence in her cover letter:

Seven years ago I resigned from a career in business to raise a family.

Her opening could easily have suggested an enterprising person had she thought to lead with her strengths:

Your organization may be interested in a seasoned personnel administrator who has started her own successful business and consulted with an executive search firm concurrently with starting a family.

Do not indulge in subjective self-praise:

The enclosed resume sketches my background and shows my creativity, outstanding leadership skills, and writing ability.

Let the employer draw inferences from factual statements! Or if you wish to point to the clear inference, be modest and less blunt. For example:

> While still a student, I have obtained useful managerial skills through various activities. For example, my work on the student newspaper helped to sharpen my business, editorial, and writing skills. During my second summer, an administrative-assistant position in a small office helped to train me in bookkeeping procedures, employee benefits, and general office systems. Recently, as a management intern assigned to the Deputy Mayor's Office, I obtained excellent exposure to public management problems.

Here is another good example:

> During my graduate school career, I have tried to combine a seriousness of intellectual purpose with the development of interpersonal leadership skills. As chairman of the Student-Faculty Committee on Student Life, president of the student body, and senator-at-large in the university senate, I have gained useful training for a career in educational administration.

Sometimes a personality trait, such as leadership, is an important criterion of the job. If it seems important to stress such a quality do so substantively, as in the following example:

> I have learned that an effective personality is an important asset in a successful public relations career. As a result, I have tried to polish and refine my interpersonal skills through a number of special workshops on leadership styles and interpersonal effectiveness. When I received the "Outstanding Student Leader" Award at Yale last year, I felt encouraged about my own future potential to work effectively with others.

Avoid the puffery of the alternative version:

> I believe that I have the kind of sparkling personality that is essential to a successful career in public relations.

A good cover letter should not exceed one page. Save the elaborate detail for the interview, when the employer is certain to be more interested. Keep the paragraphs in your letter short for the greater visual ease of your reader.

Center the letter, and allow two to four inches as a border under your signature. The letter should not look crowded. Top and side margins should be generous.

Use bond paper in 8½″ × 11″ size. Ideally, the paper should match that of your resume, but if that proves too difficult, don't worry about it. Avoid corrasable paper unless you are such a poor typist that *not* using it will make it impossible for you to type letters. You might consider paying someone else to type the letter on bond for you. If at all possible, don't use an old college typewriter for your letters. As with your resume, an IBM Selectric or Executive typewriter will enhance your text.

Neatness is important. If your cover letters have obvious corrections, there may be adverse reactions. In any event you'll look unprofessional.

Proofread carefully for errors. A mispelled name, a run-on sentence, or a typographical error can easily cost you an interview! You cannot be too much of a perfectionist in this area.

If you plan to send out a hundred resumes to one specific kind of employer, convenience is important. Write one good cover letter and reproduce it as follows.

Have just the body of a form letter (e.g., no employer name or salutation) typed on an excellent typewriter. Then have it photo-offset in the desired quantities. Using the *original* typewriter, type in the necessary individual name and salutation. Then have each letter photocopied onto bond paper for about five cents each. No one will detect that the product is not an individual letter if your photocopy is on bond paper. A second, less complicated alternative is to follow precisely the same procedure but to stop at the photocopy step. Instead have your letters photo-offset onto bond paper. The disadvantage is that the employer will be able to see that you have used a form letter.

Do not type new details into the actual body of a reproduced letter; proper alignment is extremely difficult to achieve. Even the slightest variation will probably be detected.

A Checklist

Want to write a cover letter? Sit down and fill in the blanks:

1. I am *(June graduate; M.B.A., from N.Y.U.)* _____
2. I want a job as *(a junior accountant)* _____
3. I can offer you *(skills, experience)* _____
4. I am enclosing a resume _____
5. I want to see you *(request an interview)* _____
6. Thank you!

Now polish the language, flesh out the concepts, and you have an effective cover letter.

Personnel officers' preferences in letters of application and resumes

BARRON WELLS, NELDA SPINKS, AND JANICE HARGRAVE

When this survey was conducted, the authors were all members of the faculty of the University of Southwestern Louisiana at Lafayette.

Introduction

The job-getting process is one application of persuasion communication. No matter how capable persons may be in certain aspects of job performance, they will never get the opportunity to put those capabilities into practice unless they can first persuade employers to hire them. Most employers have a choice of prospective employees from which to choose, so the job-seekers must present their qualifications in the best way possible if they are going to persuade the employer to consider their application beyond an initial cursory scan. Furthermore, most employers recognize that those same communications skills that enable applicants to present their qualifications in a clear and vivid way will also enable them to perform many important on-the-job tasks. Then how can prospective employees present themselves and their qualifications in the best possible way?

The procedure

Faculty members at the University of Southwestern Louisiana conducted a survey of the 500 largest corporations in the United States as listed in the *Fortune* directory, 1977. This survey asked the chief personnel officers of those corporations for their preferences concerning content, appearance, and length

Reprinted from the *ABCA Bulletin*, XLIV (June, 1981) by permission of Barron Wells, Nelda Spinks, and Janice Hargrave. Full title: "A Survey of the Chief Personnel Officers in the 500 Largest Corporations in the United States to Determine Their Preferences in Job Application Letters and Personal Resumes."

of cover letters and resumes. Of the 500 opinionnaires that were mailed out, 175 were returned—a 35 percent response. To a series of statements about initial contact by job applicants, cover letters, and resumes, the respondents were asked to indicate:

1. strong agreement,

2. moderate agreement,

3. undecided,

4. moderate disagreement, or

5. strong disagreement,

A. meaning not clear,

B. no response intended.

The respondents were then asked to indicate which college courses they believed would be of considerable value to prospective job applicants in preparing letters of application and resumes. Finally, they were asked to indicate the major problems encountered by personnel directors involving application letters and resumes, and the major trends for the future they could foresee in the content of job application letters and resumes.

The responses of the corporate chief personnel officers are as follows:

QUESTION	ANSWER CHOICES:	PERCENT OF RESPONDENTS MARKING EACH ANSWER CHOICE						
		1	2	3	4	5	A	B
INITIAL CONTACT								
1. The initial contact with a job applicant should be a personal interview...........		8	14	3	50	21	3	1
2. Written information about the applicant is preferred for the initial contact with the job applicant.....................		55	38	2	4	—	1	—
3. If the initial contact with the job applicant is to be in writing, a resume *only* is preferred............................		10	22	9	41	17	1	—
4. Both a cover letter and a resume must be obtained from a job applicant for a personnel director to have all the necessary information............................		26	35	7	18	11	1	1
5. A letter of recommendation concerning the applicant is enough information for hiring a person.......................		—	1	—	14	85	—	1
6. The cover letter and resume should be typewritten........................		58	30	2	8	—	—	1

		PERCENT OF RESPONDENTS MARKING EACH ANSWER CHOICE						
QUESTION	ANSWER CHOICES:	1	2	3	4	5	A	B

INITIAL CONTACT, Continued

	1	2	3	4	5	A	B
7. Handwritten cover letters and resumes are acceptable	4	31	14	23	26	—	1
8. Cover letters and resumes may be either handwritten or typewritten; there is no preference	6	21	6	34	29	—	5
9. Graphoanalysis is used on handwritten letters and resumes to match the right persons to the right job	1	—	2	11	75	3	9
10. With the use of graphoanalysis, handwritten letters and resumes can reveal characteristics of the writers	2	5	16	12	37	3	25

COVER LETTER

	1	2	3	4	5	A	B
1. A cover letter *only* reveals all the needed information about a job applicant	—	1	3	26	67	2	1
2. Letters of application are welcomed even though there are no job openings at present	36	53	3	3	3	1	—
3. Letters of application should include a reason why the job applicant is interested in *this* job	35	46	10	7	1	1	—
4. An attention-getting first sentence will stimulate interest in a particular job applicant's letter of application	11	30	27	24	7	1	1
5. The tone of the letter of application is important	34	56	4	4	1	—	1
6. Good grammar and spelling are essential in the letter of application	80	17	2	—	—	—	1
7. A job applicant should state his understanding of the requirements of the position in his letter of application	13	40	19	21	4	2	1
8. An application letter should show how the job applicant's education and experience fit the job requirements	32	46	10	10	3	—	—
9. A potential employee should ask for a personal interview in his letter of application	20	49	18	10	3	—	—

RESUME

	1	2	3	4	5	A	B
1. A resume of one page is preferred	29	38	17	10	1	1	5
2. Two-page resumes are preferable and cover the person's qualities well	10	26	23	26	9	1	6
3. A three-page resume is the type most preferred	2	2	11	32	49	—	5

QUESTION	ANSWER CHOICES:	PERCENT OF RESPONDENTS MARKING EACH ANSWER CHOICE						
		1	2	3	4	5	A	B
RESUME, Continued								
4. Keeping a resume neat is essential ..		71	25	1	1	2	—	1
5. A photograph included with the cover letter and resume is desirable............		3	17	10	25	41	1	3
6. Social aspects concerning the job applicant should be listed in the resume......		3	26	15	28	14	13	1
7. Commendations of the job applicant should be listed in the resume		10	43	17	18	9	3	1
8. One reference is adequate in the resume		2	8	18	41	20	1	11
9. Two references are adequate in the resume		2	23	19	29	16	1	11
10. Three references are adequate in the resume................................		7	38	15	15	15	1	10
11. More than three references should be included in the resume		—	5	14	35	34	1	11
12. The types of persons listed as references in the resume are important		13	41	9	16	13	3	4
13. The prospective employee should bring a copy of his resume to the job interview................................		61	36	—	1	1	—	—
14. Military service of the applicant should be included in the resume...............		42	45	6	4	—	3	1
15. A complete transcript of college grades should be attached to the resume ...		7	31	20	32	11	—	—
16. Personal information such as date of birth, phone, address, marital status, dependents, etc., should be included in the resume.		28	33	11	7	12	3	6
17. A resume should contain an applicant's physical and health status..........		11	34	18	20	10	1	6
18. General as well as specific educational qualifications—such as majors, minors, and degrees—should be included in the resume.		54	41	2	2	—	1	—
19. Willingness to relocate should be included in the resume		55	36	4	3	1	—	1
20. A list of scholarships, awards, and honors should be included in the resume. ...		39	51	6	3	1	—	—
21. The resume should contain previous work experience concerning jobs held, dates of employment, company addresses, and reasons for leaving		74	24	1	2	—	—	—

		PERCENT OF RESPONDENTS MARKING EACH ANSWER CHOICE						
QUESTION	ANSWER CHOICES:	1	2	3	4	5	A	B
RESUME, Continued								
22. Special aptitudes should be listed in the resume		30	50	12	5	1	1	1
23. A list of grades in major or minor subjects in college should be included in the resume		15	26	18	30	10	—	1
24. A resume should list the high school attended, class rank, and date of graduation of the job applicant		13	28	19	28	10	1	1
25. The major source of a person's financing while in college should be contained in the resume		10	42	24	14	9	1	1
26. A resume should contain the salary requirements of a job applicant		20	37	18	14	10	—	1
27. Social data, such as fraternities, athletics, clubs, and sororities should be listed in the resume		11	37	26	19	4	1	2
28. The traditional order in which information is presented in a resume is desirable (personal information, education, experience, references)		23	48	10	13	3	1	2
29. An applicant's experience should be listed first in the resume		7	12	26	40	5	1	8
30. The education of the applicant should be listed first in the resume		7	25	25	28	4	2	9
31. The first listing on a resume should be the names of personal references		1	2	8	29	52	1	7
32. The applicant's strongest points (education, work experience, etc.) should be listed first in the resume, without regard to any rule for presentation		29	20	14	23	8	1	6

Summary

The study yielded many useful pieces of information concerning initial contact by job seekers, job application letters, and resumes.

Initial contact

The initial contact by the job applicant should be in writing, and letters of application are highly desirable even though there are no job openings at present in the company.

Letters of application

Tone, good grammar, proper spelling, and neatness are essential in the letter of application and in the resume.

In application letters, applicants should show how their qualifications fit the job requirements, and then ask for an interview.

Resume

A one- or two-page resume is much more desirable than a three-page resume. A photograph is not essential, but commendations of the job applicant should be included.

Three references should be listed, and the types of reference persons listed are of major importance to the personnel director in the company.

High school grades, rank, and date of graduation are not necessary.

A resume should definitely contain the personal information about an applicant: health status, military status, and willingness to relocate.

Previous work experience and/or special aptitudes are of major importance in the resume. Also, the major source of financing during college should be listed in an applicant's resume, and the traditional order of listing information in a resume is preferred over all other methods.

College courses

The four most valuable courses to prospective job-seekers in writing letters of application and resume are: Business Letter Writing and Communication, Written and Spoken English, English Composition, and Technical Writing.

Problems encountered

The major problems with application letters and resumes were that job applicants fail to list career objectives and specific job objectives, both of which give a clearer view of their plans, goals, and aspirations for the future.

Many personnel directors felt that job applicants had no knowledge of the company to which they are applying for work.

One very distinct concern of personnel directors was that they should know the applicant's feelings concerning relocation and travel.

Many applicants tended to "oversell" themselves and to be vague instead of specific about themselves.

The future

The trend now is to put professionalism into preparing application letters and resumes so that they have a "marketing" (selling) approach.

Some personnel directors indicated that they would like application letters and resumes to be precise, with the use of key wording, and that application letters and resumes should be personalized—not use a form approach.

Job applicants should stress to a personnel director their value to the company and should let the company know how much they need them.

Personnel directors indicated that in the future, resumes should stress validated skills rather than experiences and that college transcripts should be included with the resume.

It is felt by personnel directors that the tighter the job market, the better are the application letters and the resumes they receive.

The general trend will be toward greater definition of skills, abilities, aptitudes, and objectives, and toward more attention to conciseness in describing experiences, education, and career objectives.

6

Annotated bibliography

*T*his annotated bibliography provides more than 275 additional sources of information on business and technical writing.

For the convenience of its users, the bibliography is divided into four sections:

1. Bibliographies

2. Books (including textbooks)

3. Anthologies

4. Journal articles.

Bibliographies

Abshire, Gary M. "Writing on Writing." *IEEE Transactions on Professional Communication*, PC-25 (December, 1982), 211–219. A simple list divided into three sections—general, business, and technical writing—of 261 books, articles, and other materials published between 1953 and 1981.

Alred, Gerald, and others. *Business and Technical Writing: An Annotated Bibliography of Books, 1880–1980*. Metuchen, NJ: Scarecrow, 1981. An annotated list of 874 books.

Balachandran, Sarojini. *Employee Communication: A Bibliography*. Urbana, IL: ABCA, 1976. An annotated list of items published between 1965 and 1975 that discuss the communication gap between managers and employees.

_____. *Technical Writing: A Bibliography*. Urbana, IL: ABCA and Washington, DC: STC, 1977. An annotated list of publications on scientific and technical writing since 1965.

Bankston, Dorothy, and others. "Bibliography of Technical Writing." *Technical Writing Teacher*, 3 (Fall 1975), 29–41; 4 (Fall 1976), 32–43; 6 (Fall 1978), 24–29. A continuing list of bibliographies, books, book reviews, articles, and technical reports; see Book and others below.

Bibliography of Publications Designed to Raise the Standard of Scientific Literature. Paris: UNESCO, 1963. An annotated list (in English) of 354 items (in English, French, Russian, Spanish, German, Bulgarian, Danish, Polish, and Czech) in six areas including general works on language and composition, technical writing, and readings in science for technical authors.

Blackman, Carolyn M. "A Bibliography for Beginning Teachers of Technical Writing." In *Technical and Professional Communication*. Thomas M. Sawyer, ed. Ann Arbor, MI: Professional Communication Press, 1977. Organizations, periodicals, conferences, meetings, institutes, bibliographies, articles, standard references, and general orientation materials on technical writing.

Book, Virginia, and others. "Bibliography of Technical Writing." *Technical Writing Teacher*, 5 (Fall 1977), 26–35; 9 (Fall 1981), 35–47; 9 (Spring 1982), 196–202; 10 (Fall 1982), 54–74. A continuing list of bibliographies, books, book reviews, articles, and technical reports; see Bankston and others above.

Bowman, Mary Ann. "Books on Business and Technical Writing in the University of Illinois Library." *Journal of Business Communication*, 12 (Winter 1975), 33–67. A list of books written in English between 1950 and 1973 whose titles appeared in the card catalogue of the University of Illinois library under a variety of subject headings related to business and technical writing.

_____, and Joan D. Stamas. *Written Communication in Business: A Selective Bibliography, 1967–1977*. Champaign, IL: ABCA, 1980. An annotated list of over 800 books and articles.

Burkett, Eva M. *Writing in Subject-Matter Fields: A Bibliographic Guide with Annotations and Writing Assignments*. Metuchen, NJ: Scarecrow, 1977. Brief individual sections on writing in law and science, technical and business writing, interdisciplinary writing, and other topics.

Burns, Shannon, and others. *An Annotated Bibliography of Texts on Writing Skills*. New York: Garland, 1976. Over 400 items on grammar and usage, composition, rhetoric, and technical writing.

Carter, Robert. *Communication in Organizations: An Annotated Bibliography and Sourcebook*. Detroit: Gale, 1972. Includes more than 75 items on business writing.

Cunningham, Donald H. "Bibliographies of Technical Writing Materials." *Technical Writing Teacher*, 1 (Winter 1974), 9–10. Instructional materials.

_____. "Books on Police Writing." *College Composition and Communication*, 23 (May, 1972), 199–201. A review of books published between 1957 and 1972 on the fundamentals of police writing.

_____, and Vivienne Hertz. "An Annotated Bibliography on the Teaching of Technical Writing." *College Composition and Communication*, 21 (May, 1970), 177–186. Sixty-seven entries; especially designed for beginning teachers of technical writing.

———. "Bibliography: Police Report Writing." *Police Chief*, 38 (August, 1971), 44, 49–50. One hundred and two items including books, sections of books, bound and unbound manuals and bulletins, articles, and materials indirectly related to police reporting.

Denton, L. W., and W. E. Rivers. "Consulting and In-House Writing Courses: A Selected Bibliography." In *Communication Training and Consulting in Business, Industry, and Government*. William J. Buchholz, ed. Urbana, IL: ABCA, 1983. An annotated list of sources for information on consulting, professional training, teaching methodology, and successful short courses in business communications.

Dorrell, Jean, and Betty Johnson. "A Comparative Analysis of Topics Covered in Twenty College-Level Communication Textbooks." *ABCA Bulletin*, 44 (September, 1982), 11–16. A breakdown by major and minor topics covered, an assessment of textbook readability using the Gunning Fog Index, and statistics on numbers of adoptions by colleges and universities.

Fearing, Bertie E., and Thomas M. Sawyer. "Speech for Technical Communicators: A Bibliography." *IEEE Transactions on Professional Communication*, PC-23 (March, 1980), 53–60. A list of 178 resources for technical communicators interested in improving their speaking skills.

Fielden, John S. "For Better Business Writing." *Harvard Business Review*, 43 (January–February, 1965), 164–166, 169–170, 172. A list of reference books designed to help people in business write better.

Goldberg, Jay J. "A Survey of Scholarly Works in Technical Writing." *Technical Communication*, 22 (1st Quarter 1975), 5–8. A discussion of 35 items dealing with readability, professional training, and information transfer.

Goldstein, Jone R., and Robert R. Donovan. *A Bibliography of Basic Texts in Technical and Scientific Writing*. Washington, DC: STC, 1982. An annotated bibliography of texts supplemented by a list of additional resource materials.

Keene, Michael L., and Merrill D. Whitburn. "Audience Analysis for Technical Writing: A Selective, Annotated Bibliography." In *Teaching Technical Writing: Teaching Audience Analysis and Adaptation*. Paul V. Anderson, ed. Morehead, KY: Association of Teachers of Technical Writing (Anthology #1), 1980. Detailed annotations of more than 25 articles and books discussing theoretical as well as practical considerations concerning audience.

Killingworth, M. J. "A Bibliography on Proposal Writing." *IEEE Transactions on Professional Communication*, PC-26 (June, 1983), 79–83. Annotations of 80 items.

Larson, Richard L. "Selected Bibliography of Research and Writing about the Teaching of Composition." *College Composition and Communication*, 26 (May, 1975), 187–195; 27 (May, 1976), 171–180; 28 (May, 1977), 181–193; 29 (May, 1978), 181–194; and 30 (May, 1979), 196–213. Some items on business and technical writing; annotated.

Leonard, Don. "An Annotated Bibliography of Recent Articles on Company Communications Training." *ABCA Bulletin*, 45 (June, 1982), 48–52. Thirty-four entries.

Lodge, Frank L. "Specifications and Catalogs Frequently Used in the Presentation of Technical Data." In *Handbook of Technical Writing Practices*. Stello Jordan and others, eds. Vol. 2. New York: Wiley, 1971. An annotated list of specifications and catalogues issued by agencies of the Defense Department.

McClure, Lucille. "Two Bibliographies: Technical Writing Books in Print. Photo-typesetting." *IEEE Transactions on Engineering Writing and Speech*, EWS-8 (December, 1965), 65–70. Fifty-nine books on technical writing.

Miller, Carolyn R., and Bertie E. Fearing. "Resources for Teachers of Business, Technical and Vocational Writing." In *Technical and Business Communication in Two-Year Programs*. W. Keats Sparrow and Nell Ann Pickett, eds. Urbana, IL: NCTE, 1983. An annotated list of organizations, institutes and workshops, journals, materials for teacher preparation, textbooks for freshmen and sophomore courses, standard reference works and resources, and bibliographies.

Oslund, R. R. "A Bibliography of Proposals." *STWP Review*, 11 (April, 1964), 13–14. Eighty-four items on proposal writing.

Philler, Theresa, and others. *An Annotated Bibliography on Technical Writing, Editing, Graphics, and Publishing, 1950–1965.* Washington, DC: Society for Technical Writers and Publishers; Pittsburgh: Carnegie Library, 1966. Two thousand books, articles, and papers.

Smith, Julian F., and others. "Style Manuals: Guides for Technical Writing." *Journal of Chemical Education,* 42 (July, 1965), 373–375. A review of 41 style manuals designed for technical writers.

Van Veen, Frederick. "An Index to 500 Papers through 1962 on Engineering Writing and Related Subjects." *IEEE Transactions on Engineering Writing and Speech,* EWS-6 (September, 1963), 50–58. Author-subject index; many of the papers listed have not been published.

Walsh, Ruth M., and Stanley J. Birkin. *Business Communication: An Annotated Bibliography.* Westport, CT: Greenwood, 1980. Sixteen hundred books, articles, reviews, and dissertations written between 1960 and 1979.

Walsh, Ruth M., and others. "Business Communications: A Selected, Annotated Bibliography." *Journal of Business Communications,* 11 (Fall, 1973), 65–112. A list of over 400 books and articles published between 1940 and 1972; emphasizes reports and business correspondence.

Walter, John A. "Basic Recommended Reference Shelf: A Selected Bibliography of Technical and Scientific Writing." In *Handbook of Technical Writing Practices.* Stello Jordan and others, eds. Vol. 2. New York: Wiley, 1971. An annotated list of reference books on grammar, style, usage, and business and technical communications.

———. "Style Manuals." In *Handbook of Technical Writing Practices.* Stello Jordan and others, eds. Vol. 2. New York: Wiley, 1971. An annotated list of more than 60 general, institutional, and industrial style manuals and authors' guides.

White, Jane F., and Patty G. Campbell. *Abstracts of Business Communication, 1900 through 1970.* Urbana, IL: ABCA, 1982. Abstracts of theses and dissertations in business communications.

Books

Adelstein, Michael E., and W. Keats Sparrow. *Business Communications.* New York: Harcourt Brace Jovanovich, 1983. A comprehensive introductory text.

Alvarez, Joseph A. *Elements of Technical Writing.* New York: Harcourt Brace Jovanovich, 1980. A handbook taking a descriptive approach to principles of grammar, punctuation, and style.

Andrews, Deborah C., and Margaret D. Blickle. *Technical Writing Principles and Forms.* 2nd ed. New York: Macmillan, 1982. A comprehensive introductory text that takes a rhetorical approach to technical writing.

Austen, David, and Tim Crosfield. *English for Nurses.* London: Longman, 1976. A basic text for students in nursing programs.

Backman, Lon F. *A Digest on the Elements of Proposal Writing.* Olympia, WA: State of Washington Planning and Community Affairs Agency, n.d. Detailed guidelines written in the form of an actual proposal.

Barnett, Marva T., and J. L. Smith. *Effective Communications for Public Safety Personnel.* New York: Delmar, 1978. An introductory text for students in police and fire-training programs.

Barrass, Robert. *Scientists Must Write.* London: Chapman and Hall, 1978. A brief guide to better writing for working scientists and engineers.

Barzun, Jacques, and Henry F. Graff. *The Modern Researcher.* 3rd ed. New York: Harcourt Brace Jovanovich, 1977. A scholarly discussion of the principles that should inform any kind of research.

Bates, Jefferson D. *Writing with Precision.* Washington, DC: Acropolis, 1978. A self-help text designed to enlist government writers in the "war" against gobbledygook in bureaucratic prose.

Bennett, John B. *Editing for Engineers.* New York: Wiley, 1970. A brief handbook for engineers who write and managers who edit technical documents.

Berenson, Conrad, and Raymond R. Colton. *Research and Report Writing for Business and Economics.* New York: Random House, 1971. A comprehensive guide for beginning college students who need to write reports and research papers.

Bingham, Earl G. *Pocketbook for Technical and Professional Writers.* Belmont, CA: Wadsworth, 1982. A brief handbook.

Blicq, Ron S. *Technically Write!* 2nd ed. Englewood Cliffs, NJ: Prentice-Hall, 1981. A basic text using a modern engineering consulting firm as the setting for each assignment.

Blumenthal, Lassor A. *The Complete Book of Personal Letter Writing and Modern Correspondence.* Garden City, NY: Doubleday, 1969. A comprehensive guide to social, business, and professional letter writing.

Bolles, Richard Nelson. *What Color Is Your Parachute?* Berkeley, CA: Ten Speed Press, updated annually. A practical guide to the job hunt and career change.

Bostwick, Burdette. *Resume Writing.* New York: Wiley, 1976. A comprehensive guide with discussions and examples of different resume formats.

Bowman, Joel P., and Bernadine P. Branchaw. *Business Report Writing.* Chicago: Dryden, 1984. A comprehensive introduction to report writing for teachers, students, and professionals.

Brand, Norman, and John O. White. *Legal Writing: The Strategy of Persuasion.* New York: St. Martin's, 1976. A text applying the principles of advanced composition and persuasion to the specific writing needs of prelaw, paralegal, and law students.

Brennan, Lawrence D., and others. *Resumes for Better Jobs.* New York: Simon and Schuster, 1973. Over 200 model resumes.

Brinegar, Bonnie Carter, and Craig Barnwell Skates. *Technical Writing: A Guide with Models.* Glenview, IL: Scott, Foresman, 1983. An introductory text/reader emphasizing four basic technical writing models: information, investigation, evaluation, and persuasion.

Brown, Harry M. *Business Report Writing.* New York: Van Nostrand, 1980. A thorough discussion of all aspects of report writing for students and professionals.

————, and Karen K. Reid. *Business Writing and Communication.* New York: Van Nostrand, 1979. An introductory text.

Brunner, Ingrid, and others. *The Technician as Writer.* Indianapolis: Bobbs-Merrill, 1980. A technical report writing text for students in junior colleges.

Brusaw, Charles T., and others. *Handbook of Technical Writing.* 2nd ed. New York: St. Martin's, 1982. An easy-to-use, thorough handbook containing over 500 alphabetically arranged entries.

Carlsen, Robert D., and Donald L. Vest. *The Encyclopedia of Business Charts.* Englewood Cliffs: NJ: Prentice-Hall, 1977. A comprehensive reference guide to business charts and their style.

Carr-Ruffino, Norma. *Writing Short Business Reports.* New York: Gregg, 1980. A text-workbook for classes or industrial training programs emphasizing report writing.

Cheatham, T. Richard, and Keith V. Erickson. *The Police Officer's Guide to Better Communication.* Glenview, IL: Scott, Foresman, 1984. Basic guidelines for communicating effectively with witnesses, suspects, other officers, and the public.

The Chicago Manual of Style. 13th ed. Chicago: University of Chicago, 1982. A standard reference for authors, editors, copywriters, and proofreaders.

Cloke, Marjane, and Robert Wallace. *The Modern Business Letter Writer's Manual.* New York: Doubleday, 1974. Handy paperback guide for writers of business letters.

Cremmins, Edward T. *The Art of Abstracting.* Philadelphia: ISI, 1982. A thorough discussion of the rules for reading and writing abstracts.

Crouch, W. George. *Bank Letters and Reports.* New York: American Institute of Banking, 1961. An introductory text for bank employees.

Damerst, William A. *Clear Technical Reports.* New York: Harcourt Brace Jovanovich, 1972. A comprehensive text emphasizing the need for clarity.

Davis, Ken. *Better Business Writing: A Process Approach.* Columbus: Merrill, 1983. An introductory text for junior college or short corporate business writing courses.

Dawe, J. *Writing Business and Economics Papers, Theses, and Dissertations.* Totowa, NJ: Littlefield, Adams, 1975. A comprehensive style manual.

Day, Robert A. *How to Write and Publish a Scientific Paper.* Philadelphia: ISI, 1979. A "cookbook" for scientists to help them put their research into publishable form.

Dirckx, John H. *Dx + Rx: A Physician's Guide to Medical Writing.* Boston: G. K. Hall, 1977. A handbook emphasizing accuracy, clarity, and readability.

Dodds, Robert H. *Writing for Technical and Business Magazines.* New York: Wiley, 1969. Practical discussions of the publishing process from the points of view of writer and editor alike.

Draughton, Clyde O. *Practical Bank Letter Writing.* Boston: Bankers Publishing Co., 1971. A guide to the kinds of letters employees of financial institutions must write.

Ehrlich, Eugene, and Daniel Murphy. *The Art of Technical Writing.* New York: Crowell, 1969. A handbook.

Eisenberg, Anne. *Reading Technical Books.* Englewood Cliffs, NJ: Prentice-Hall, 1978. A basic junior-college text designed to show students how to find and organize technical information.

Elbow, Peter. *Writing without Teachers.* New York: Oxford, 1973. A brief introduction to the writing process.

Emerson, Lynn A. *How to Prepare Training Manuals.* Albany: University of the State of New York, Department of Education, 1952. Ten chapters following the order in which the steps of manual writing should be carried out.

Enrick, Norbert L. *Handbook of Effective Graphic and Tabular Communication.* Huntington, NY: Krieger, 1980. A thorough discussion supplemented by numerous examples.

Fear, David E. *Technical Communication.* 2nd ed. Glenview, IL: Scott, Foresman, 1981. A rhetoric and handbook.

Flesch, Rudolf. *The ABC of Style.* New York: Harper and Row, 1980. A reference guide explaining how to avoid the pedantic, the trite, the pompous, and the verbose in writing.

_____. *The Art of Readable Writing.* Rev ed. New York: Harper and Row, 1973. A general discussion based on the premise that one should as much as possible write the way one talks; includes a discussion of the Flesch Readability Formula.

_____. *How to Write Plain English.* New York: Harper and Row, 1981. A guide for consumers and lawyers.

Fluegelman, Andrew, and J. J. Hewes. *Writing in the Computer Age.* Garden City, NY: Anchor, 1983. A guide to computers that explains how to use a word processor to write and edit.

French, Christopher, and others, eds. *The Associated Press Style Book and Libel Manual.* Reading, MA: Addison-Wesley, 1980. The working journalist's bible covering the rules of grammar, punctuation, and usage.

Gallagher, William J. *Writing the Business and Technical Report.* Boston: CBI, 1981. A detailed, process approach to report writing.

Garrison, Roger. *How a Writer Works.* New York: Harper and Row, 1981. A concise guide for professionals that emphasizes revision as the key to effective writing.

Gelderman, Carol. *Better Writing for Professionals.* Glenview, IL: Scott, Foresman, 1983. A concise guide and reference offering advice on writing and publishing articles for career advancement.

Goswami, D., and others. *Writing in the Professions.* Washington, DC: American Institutes for Research, 1981. A systematic, process-oriented resource book.

Gunning, Robert. *New Guide to More Effective Writing in Business and Industry.* Boston: Industrial Education Institute, 1963. A quick reference designed for the technician on the job.

_____. *The Technique of Clear Writing.* Rev. ed. New York: McGraw-Hill, 1968. A detailed discussion of the Fog Index and selected principles of clear writing.

Hafer, W. Keith, and Gordon E. White. *Advertising Writing.* St. Paul: West, 1977. A comprehensive text on public relations and advertising writing.

Harris, John S., and Reed H. Blake. *Technical Writing for Social Scientists.* Chicago: Nelson Hall, 1976. A rhetoric for social scientists and students in social science programs.

Hart, Andrew W., and James A. Reinking. *Writing for Career-Education Students.* 2nd ed. New York: St. Martin's Press, 1982. A comprehensive introductory text for students in vocational and technical programs.

Haughney, John. *Effective Catalogs.* New York: Wiley, 1968. Procedures for producing a basic marketing tool, the company catalog.

Henderson, Greta L., and Price R. Voiles. *Business English Essentials.* 6th ed. New York: Gregg, 1980. A basic text-workbook.

Herbert, A. J. *The Structure of Technical English.* London: Longman, 1965. A grammar book designed to give foreign technicians and engineers practice in mastering the skills they need to communicate effectively in English.

Hill, L. Brooks. *The Military Officer's Guide to Better Communication.* Glenview, IL: Scott, Foresman, 1984. A brief guide designed to show military personnel how to improve their effectiveness as leaders and managers through better oral and written communication skills.

Himstreet, William C., and Wayne M. Baty. *Business Communications.* 7th ed. Boston: Kent, 1984. A comprehensive introductory text.

Hirschhorn, Howard H. *Writing for Science, Industry, and Technology.* New York: Van Nostrand, 1980. An introductory text for students majoring in the physical or social sciences, business, and technological areas.

Holcombe, Marya W., and Judith K. Stein. *Writing for Decision Makers: Memos and Reports with a Competitive Edge.* Belmont, CA: Lifetime Learning Publications, 1981. An advanced text stressing the writing process.

Holtz, Herman, and Terry Schmidt. *The Winning Proposal: How to Write It.* New York: McGraw-Hill, 1981. A thorough overview of the proposal process offering valuable insights into the way government works.

Hoover, Hardy. *Essentials for the Technical Writer.* Rev. ed. New York: Dover, 1981. A brief guide to writing reports and specifications.

Houp, Kenneth, and Thomas Pearsall. *Reporting Technical Information.* 5th ed. New York: Macmillan, 1984. Well-illustrated text for advanced writing courses.

Howard, C. Jeriel, and David A. Gill. *Desk Copy: Modern Business Communications.* San Francisco: Canfield, 1971. A basic text for students in business and secretarial programs.

Huseman, Richard C., and others. *Business Communication, Strategies and Skills.* Hinsdale, IL: Dryden, 1981. A basic text in communication theory and its practical application to the world of business.

Jackson, Tom. *The Perfect Resume.* Garden City, NY: Anchor, 1980. A guide to cover letters, resumes, and job-finding, and salary-negotiating strategies supplemented by 55 sample resumes.

Jacobi, Ernest. *Work at Writing.* Rochelle Park, NJ: Hayden, 1980. A workbook to accompany *Writing at Work;* see below.

―――. *Writing at Work.* Rochelle Park, NJ: Hayden, 1976. A text with emphasis on general principles rather than mechanics; intended for writers on the job.

Janis, J. Harold. *The Business Research Paper.* New York: Hobbs, Dorman, 1967. A compact guide to report formats and mechanics.

―――. *Writing and Communicating in Business.* 3rd ed. New York: Macmillan, 1978. A comprehensive text for beginning or advanced courses.

―――, and Howard R. Dressner. *Business Writing.* 2nd ed. New York: Barnes and Noble, 1972. A self-help guide to letter and report writing.

Jaquish, Michael P. *Personal Resume Preparation.* New York: Wiley, 1968. Step-by-step instructions for the preparation of resumes.

Keithley, Erwin M., and Philip J. Schreiner. *A Manual of Style for the Preparation of Papers and Reports.* 3rd ed. Cincinnati: South-Western, 1980. A style guide.

Keithley, Erwin M., and Margaret H. Thompson. *English for Modern Business.* 4th ed. Homewood, IL: Irwin, 1982. A basic text for a refresher course in the fundamentals of grammar and writing.

Kett, Merriellyn, and Virginia Underwood. *How to Avoid Sexism.* Chicago: Lawrence Ragan Communications, 1978. A guide for writers, editors, and publishers that offers detailed discussions of sexism and the law, the generic "he," problems with forms of address, sex-role stereotyping, and acceptable practices for writing about both men and women.

Kleppner, Otto, and Thomas Russell. *Advertising Procedure.* 8th ed. Englewood Cliffs, NJ: Prentice-Hall, 1983. The standard text on all aspects of advertising.

Kolin, Philip C. *Successful Writing at Work*. Lexington, MA: Heath, 1982. An introductory business writing text emphasizing product rather than process.

_____, and Janeen L. Kolin. *Professional Writing for Nurses in Education, Practice, and Research*. St. Louis: Mosby, 1980. A technical writing text for nurses at all professional levels.

Krey, Isabelle A., and Bernadette V. Metzler. *Principles and Techniques of Effective Business Communication*. New York: Harcourt Brace Jovanovich, 1976. An introductory text-workbook.

Krupa, Gene. *Situational Writing*. Belmont, CA: Wadsworth, 1982. A basic text built around a sequence of writing situations.

Lanham, Richard A. *Revising Business Prose*. New York: Scribner's, 1981. A brief handbook and style manual.

Lannon, John M. *Technical Writing*. 2nd ed. Boston: Little, Brown, 1982. A text offering a comprehensive introduction to technical writing.

Laster, Ann A., and Nell Ann Pickett. *Occupational English*. 3rd ed. New York: Harper and Row, 1981. A text for junior-college students that emphasizes the writing process.

Lay, Mary K. *Strategies for Technical Writing*. New York: Holt, 1982. An introductory rhetoric with readings.

Lefferts, Robert. *Getting a Grant in the 80's*. 2nd ed. Englewood Cliffs, NJ: Prentice-Hall, 1982. A primer for preparing research proposals directed at corporations and foundations.

Leonard, Donald J. *Shurter's Communication in Business*. New York: McGraw-Hill, 1979. An introductory text for undergraduates or those already in business; a revision of the third edition of Robert Shurter's *Written Communication in Business* (1971).

Lesikar, Raymond V. *Basic Business Communication*. Rev. ed. Homewood, IL: Irwin, 1982. A comprehensive introductory text emphasizing practical applications rather than theory.

_____. *Business Communication, Theory and Application*. 4th ed. Homewood, IL: Irwin, 1980. A comprehensive advanced text.

_____. *Report Writing for Business*. 6th ed. Homewood, IL: Irwin, 1981. A comprehensive discussion of the process of organizing and writing business reports.

Levine, Norman. *Technical Writing*. New York: Harper and Row, 1979. A brief introductory text.

Locke, Lawrence F., and W. W. Spirduso. *Proposals That Work*. New York: Teachers College Press, 1976. Guidelines for graduate students in the sciences.

Lodwig, Richard. *Career English*. Rochelle Park, NJ: Hayden, 1981. A basic text offering vocational guidance as well as instruction in writing; includes a series of short interviews in which people talk about why they enjoy their jobs.

Londo, Richard J. *Common Sense in Business Writing*. New York: Macmillan, 1982. A text stressing the writing process and emphasizing the basic essentials of composition and clarity in writing.

Maizell, Robert E., and others. *Abstracting Scientific and Technical Literature*. New York: Wiley, 1971. Tips for beginning abstractors, information scientists, and managers.

Mandel, Siegfried, and David L. Caldwell. *Proposal and Inquiry Writing*. New York: Macmillan, 1962. An example-filled text that takes the reader through the logical steps of initial inquiry, proposal preparation, and contract.

Mehaffey, Robert E. *Writing for the Real World*. Glenview, IL: Scott, Foresman, 1980. A text designed to prepare beginning students for the writing demands of the working world.

Messer, Ronald K. *Style in Technical Writing*. Glenview, IL: Scott, Foresman, 1982. A text-workbook for vocational students.

Meyer, Harold E. *Lifetime Encyclopedia of Letters*. Englewood Cliffs, NJ: Prentice-Hall, 1983. Guidelines for writing over 500 different kinds of business and personal letters.

Miller, Casey, and Kate Swift. *The Handbook of Nonsexist Writing*. New York: Barnes and Noble, 1980. Guidelines for eliminating sexism from oral and written communication.

Mills, Gordon H., and John A. Walter. *Technical Writing*. 4th ed. New York: Holt, 1978. A comprehensive rhetorically based text.

Mitchell, John N. *Writing for Professional and Technical Journals*. New York: Wiley, 1968. A discussion of the characteristics of and style guides for professional and technical articles.

Monroe, Judson. *Effective Research and Report Writing in Government.* New York: McGraw-Hill, 1980. A self-help text for staff specialists and researchers in public agencies.

————, and others. *The Science of Scientific Writing.* Dubuque: Kendall/Hunt, 1977. A brief guide designed to help students and laboratory researchers write term papers, dissertations, and journal articles.

Moyer, Ruth, and others. *The Research and Report Handbook for Business, Industry, and Government.* Student ed. New York: Wiley, 1981. An advanced text discussing reports, proposals, policies, legal briefs, memos, letters, and minutes.

Mullins, Carolyn J. *The Complete Writing Guide to Preparing Reports, Proposals, Memos, etc.* Englewood Cliffs, NJ: Prentice-Hall, 1980. A guide for on-the-job reports.

Munter, Mary. *Guide to Managerial Communication.* Englewood Cliffs, NJ: Prentice-Hall, 1982. A brief handbook on written and oral communications.

Murphy, Herta A., and Charles Peck. *Effective Business Communication.* 3rd ed. New York: McGraw-Hill, 1980. A comprehensive text for advanced or beginning courses.

O'Hayre, John. *Gobbledygook Has Gotta Go.* Washington, DC: GPO, 1966. Suggestions about how to write direct and uncluttered reports, memos, and letters.

Oliu, Walter E., and others. *Writing That Works.* 2nd ed. New York: St. Martin's, 1984. An introductory text with a brief handbook.

Pauley, Steven E. *Technical Report Writing Today.* 2nd ed. Boston: Houghton Mifflin, 1979. A brief introductory text asking students to write on topics in their specialized technical fields for an uninformed audience—the instructor.

Pearsall, Thomas. *Audience Analysis for Technical Writing.* Beverly Hills, CA: Glencoe, 1969. A pioneering study delineating audiences in terms of their technical expertise.

————. *Teaching Technical Writing: Methods for College English Teachers.* Washington, DC: STC, 1975. A short booklet for composition teachers interested in teaching more advanced or specialized courses in technical writing.

————, and Donald H. Cunningham. *How to Write for the World of Work.* 2nd ed. New York: Holt, 1982. An introductory text especially helpful in courses where the teacher is new to the fields of technical or business writing.

Pickett, Nell Ann, and Ann A. Laster. *Technical English: Writing, Reading, and Speaking.* 4th ed. New York: Harper and Row, 1983. A comprehensive introductory text-workbook for students at two-year colleges.

Rathbone, Robert R. *Communicating Technical Information.* Reading, MA: Addison-Wesley, 1966. A self-help text for engineers and scientists who want to write more effective reports, abstracts, and journal articles.

Reid, James M., Jr., and Robert M. Wendlinger. *Effective Letters.* 3rd ed. New York: McGraw-Hill, 1978. A self-help text prepared in collaboration with the New York Life Insurance Company.

Roberts, Louise A. *How to Write for Business.* New York: Harper and Row, 1978. A brief introductory text.

Roman, Kenneth, and Joel Raphaelson. *Writing That Works.* New York: Harper and Row, 1981. Advice from two advertising executives on how to write letters, memos, reports, and fund-raising literature.

Sharf, Barbara F. *The Physician's Guide to Better Communication.* Glenview, IL: Scott, Foresman, 1984. A brief guide examining and offering solutions to the problems physicians face in communicating with patients and colleagues.

Sherman, Theodore A., and Simon S. Johnson. *Modern Technical Writing.* 4th ed. Englewood Cliffs, NJ: Prentice-Hall, 1983. A comprehensive introductory text emphasizing letter and report writing.

Shurter, Robert. *Effective Letters in Business.* 2nd ed. New York: McGraw-Hill, 1954. Popular paperback that examines the psychological principles that lie behind effective letters.

Sigband, Norman. *Communication for Management and Business.* 3rd ed. Glenview, IL: Scott, Foresman, 1982. A comprehensive introductory text supplemented by 15 reading selections.

Sorrels, Bobbye D. *The Nonsexist Communicator*. Englewood Cliffs, NJ: Prentice-Hall, 1983. Solutions to problems with gender and awkwardness in business, technical, and other kinds of writing.

Souther, James W., and Myron L. White. *Technical Report Writing*. 2nd ed. New York: Wiley, 1977. A discussion of the writing process and its application to scientific and technical report writing.

Steward, Joyce S., and Marjorie Smelstor. *Writing in the Social Sciences*. Glenview, IL: Scott, Foresman, 1983. A reader/text designed for students in the fields of psychology, sociology, anthropology, history, and economics.

Taylor, K. Phillip, and others. *Communication Strategies for Trial Attorneys*. Glenview, IL: Scott, Foresman, 1984. A brief guide offering techniques for persuading judges and juries.

Tichy, H. J. *Effective Writing for Engineers, Managers, and Scientists*. New York: Wiley, 1967. A comprehensive text that includes a brief discussion of the writing process.

Trelease, Sam F. *How to Write Scientific and Technical Papers*. Cambridge, MA: MIT Press, 1969. A handy desk manual for technical writers.

Trimble, John R. *Writing with Style, Conversations on the Art of Writing*. Englewood Cliffs, NJ: Prentice-Hall, 1975. A writer's "survival kit" blending theory with practical advice.

Turner, Rufus P. *Grammar Review for Technical Writers*. 1971; rpt. Melbourne, FL: Krieger, 1981. A brief review of practical English grammar for technicians.

Ulman, Joseph N., and Jay R. Gould. *Technical Report Writing*. 3rd ed. New York: Holt, 1972. Comprehensive text and reference for technical report writers.

U.S. Government Correspondence Manual. Washington, DC: GSA, 1977. Standards for the appearance and style of government correspondence.

U.S. Government Printing Office Style Manual. Rev. ed. Washington, DC: GPO, 1973. The official government style guide.

Visco, Louis J. *The Manager as an Editor*. Boston: CBI, 1981. Guidelines for managers on how to issue clear instructions, edit effectively, and educate subordinates to improve their writing without undermining their morale.

Warner, Joan E. *Business English for Careers*. Reston, VA: Reston, 1981. A remedial text.

Weeks, Francis W., and Kitty O. Lockner. *Business Writing Cases and Problems*. Champaign, IL: Stipes, 1980. Ninety-six cases.

Weisman, Herman W. *Technical Correspondence*. New York: Wiley, 1968. A handbook.

———. *Technical Report Writing*. 2nd ed. Columbus: Merrill, 1975. A brief report writing handbook.

Weiss, Edmond H. *The Writing System for Engineers and Scientists*. Englewood Cliffs, NJ: Prentice-Hall, 1982. A comprehensive introduction to the writing process.

Whalen, Doris H. *Handbook for Business Writers*. New York: Harcourt Brace Jovanovich, 1978. A ready reference manual discussing the forms of business writing.

———. *Handbook of Business English*. New York: Harcourt Brace Jovanovich, 1980. A guide to style, grammar, spelling, word usage, and sentence mechanics.

Wilkinson, C. W., and others. *Communicating through Letters and Reports*. 8th ed. Homewood, IL: Irwin, 1983. A comprehensive introductory text.

Williams, Joseph M. *Style: Ten Lessons in Clarity and Grace*. Glenview, IL: Scott, Foresman, 1981. Advice on quick and effective communication.

Woodford, F. Peter., ed. *Scientific Writing for Graduate Students*. New York: Rockefeller University Press, 1968. A comprehensive guide to scientific writing prepared by the Council of Biology Editors.

Wyld, Lionel D. *Preparing Effective Reports*. New York: Odyssey, 1967. A brief handbook.

Zinsser, William. *On Writing Well*. 2nd ed. New York: Harper and Row, 1980. A brief, handy reference guide.

———. *Writing with a Word Processor*. New York: Harper and Row, 1983. A step-by-step "confession" of Zinsser's encounters with his IBM Displaywriter as he wrote this book.

Anthologies

Anderson, Paul V., ed. *Teaching Technical Writing: Teaching Audience Analysis and Adaptation.* Morehead, KY: Association of Teachers of Technical Writing (Anthology #1), 1980. Four original essays and a selective annotated bibliography that provide materials for teaching audience analysis and adaptation.

————, and others, eds. *New Essays in Technical and Scientific Communication: Research, Theory, Practice.* Farmingdale, NY: Baywood, 1983. Twelve essays discussing readability and important pedagogical concerns for teachers and practitioners of technical and scientific writing.

Anderson, W. Steve, and Don Richard Cox, eds. *The Technical Reader.* 2nd ed. New York: Holt, 1984. Eighty introductory readings in technical, business, and scientific communication.

Armour, Richard, ed. *How to Write Better.* Boston: Christian Science Publishing Society, 1977. Reprints of articles on writing that appeared in the *Monitor* together with illustrative articles on a wide variety of subjects.

Bovee, Courtland, ed. *Business Writing Workshop.* Sherman Oaks, CA: Roxbury, 1980. Reprints of 41 articles discussing the forms of business writing.

Buchholz, William J., ed. *Communication Training and Consulting in Business, Industry, and Government.* Urbana, IL: ABCA, 1983. Twenty-nine essays, some previously published, and an annotated bibliography.

Cunningham, Donald H., and Herman A. Estrin, eds. *The Teaching of Technical Writing.* Urbana, IL: NCTE, 1975. Twenty-four previously published essays addressing the central pedagogical concerns of technical and scientific writing.

Douglas, George H., ed. *The Teaching of Business Communication.* Champaign, IL: ABCA, 1978. Reprints of 40 essays that appeared from 1972 to 1977 in the *ABCA Bulletin.*

Effective Communication for Engineers. New York: McGraw-Hill, 1974. Reprints of more than 50 articles from *Chemical Engineering* that discuss a wide range of topics in oral and written technical communication.

Estrin, Herman A., ed. *Technical and Professional Writing, A Practical Anthology.* 1963; rpt. New York: Preston, 1976. More than 40 articles on important aspects of business and technical writing.

Gieselman, Robert D., ed. *Readings in Business Communication.* Champaign, IL: Stipes, 1974. Twenty-two essays on a broad range of topics related to business communication.

Gould, Jay R., ed. *Directions in Technical Writing and Communication.* Farmingdale, NY: Baywood, 1978. Fourteen essays, all previously published in the *Journal of Technical Writing and Communication,* that discuss the major forms of technical writing.

Halpern, Jeanne W., ed. *Teaching Business Writing.* Urbana, IL: ABCA, 1983. Thirteen essays discussing approaches, plans, pedagogy, and research.

Hildebrandt, Herbert W., ed. *International Business Communication: Theory, Practice, Teaching throughout the World.* Ann Arbor, MI: Division of Research, Graduate School of Business Administration, University of Michigan, 1981. Fourteen essays examining business communication and marketing concepts in Europe, the Far East, North and South America, and Africa.

Huseman, Richard C., and others, eds. *Readings in Business Communication.* Hinsdale, IL: Dryden, 1981. Reprints of 36 articles discussing theoretical considerations and strategies for written and oral business communications.

Jordan, Stello, and others, eds. *Handbook of Technical Writing Practices.* 2 vols. New York: Wiley, 1971. Thirty-two essays offering a complete guide to technical writing and its various support services.

Journet, Debra, and Julie Lepick Kling, eds. *Readings for Technical Writers.* Glenview, IL: Scott, Foresman, 1984. Thirty-six selections written by professional technical writers and drawn from on-the-job situations in government, business, and industry.

Leonard, David C., and Peter J. McGuire, eds. *Readings in Technical Writing.* New York: Macmillan, 1983. Reprints of 21 articles and 5 samples that address the issues of style, audience, arrangement, and readability.

[Mann, Gerald A., and Harold E. Osborne, eds.] *Proposals and Their Preparation.* STC Anthology Series No. 1. Washington, DC: STC, 1973. Sixteen essays, all previously published by the Society for Technical Communication, discussing procedures and guidelines for writing proposals.

Nilsen, Alleen Pace, and others, eds. *Sexism and Language.* Urbana, IL: NCTE, 1977. Eighteen essays on sexism as a social, educational, and linguistic issue; an appendix reprints the NCTE Guidelines for Nonsexist Use of Language.

Pickens, Judy E., and others, eds. *Without Bias: A Guidebook for Nondiscriminatory Communication.* San Francisco: International Association of Business Communicators, 1977. Six essays and a case study that set guidelines for avoiding discrimination on the basis of race, ethnic bias, sex, or physical handicap in written communication, visual media, and meetings, conferences and workshops.

Richards, Jack C., ed. *Teaching English for Science and Technology.* SEAMO Regional English Language Centre Anthology Series No. 2. Singapore: Singapore University Press, 1976. Thirteen essays on the differences among scientific, technological, and technical English and the special problems they create for non-native speakers of English.

Rivers, William E., ed. *Business Reports: Samples from the "Real World."* Englewood Cliffs, NJ: Prentice-Hall, 1981. Thirty-eight sample reports written for various companies.

Sawyer, Thomas M., ed. *Technical and Professional Communication.* Ann Arbor, MI: Professional Communication Press, 1977. Twenty pedagogical essays.

Shaw, James, ed. *Teaching Technical Writing and Editing—In-House Programs That Work.* STC Anthology Series No. 5. Washington, DC: STC. 1976. Twelve essays, all previously published by the Society for Technical Communication, discussing problems of designing and teaching in-house technical writing courses.

Siegel, Muffy E. A., and Toby Olson, eds. *Writing Talks.* Upper Montclair, NJ: Boynton/Cook, 1983. Twelve original essays by authors with divergent backgrounds (a poet, a philosopher, a social worker, a psychologist, the director of a writing program, an EPA official, an international banker, and several college professors) who discuss problems in the teaching of writing.

Sparrow, W. Keats, and Donald H. Cunningham, eds. *The Practical Craft: Readings for Business and Technical Writers.* Boston: Houghton Mifflin, 1978. Twenty-eight essays.

Sparrow, W. Keats, and Nell Ann Pickett, eds. *Technical and Business Communication in Two-Year Programs.* Urbana, IL: NCTE, 1983. Twenty-five essays addressed directly to teachers of business and technical communication courses at junior and community colleges.

Stevenson, Dwight W., ed. *Courses, Components, and Exercises in Technical Communication.* Urbana, IL: NCTE, 1981. Twenty-one essays addressing a variety of pedagogical concerns.

Stratton, Charles R., ed. *Teaching Technical Writing: Cassette Grading.* Morehead, KY: Association of Teachers of Technical Writing (Anthology #2), 1979. Six previously published essays with updated postscripts by their authors and a brief bibliography.

Teaching Scientific Writing. English Journal, 67 (April 1978). Seventeen articles.

Vervalin, Charles H., ed. *Communication and the Technical Professional.* Houston: Gulf, 1981. Twenty-six essays discussing communication dynamics and skills for engineers and scientists.

Weeks, Francis W., ed. *Readings in Communication from Fortune.* New York: Holt, 1961. Reprints of 28 articles from *Fortune.*

Whitburn, Merrill, ed. *Teaching Technical Writing: The First Day in the Technical Writing Course.* Morehead, KY: Association for Teachers of Technical Writing (Anthology #3), 1980. Five essays discussing writing assignments.

Wilkinson, C. W., and others, eds. *Writing for Business, Selected Articles on Business Communication.* 3rd ed. Homewood, IL: Irwin, 1960. More than seventy articles on the theory and practice of business communication.

Zook, Lola M., ed. *Technical Editing: Principles and Practices.* STC Anthology Series No. 4. Washington, DC: STC, 1975. Reprints of 18 papers and journal articles.

Journal articles

Adams, Tom. "Developing a Meeting Memo." *Supervisory Management,* 25 (July, 1980), 39–42. The 13 essentials of an effective meeting memo.

Ammannito, Theresa A. "The Introduction." *STWP Review,* 9 (July, 1962), 11–14. An examination of the correlation between reader expertise and the length and complexity of introductions to technical reports and articles.

Arnold, Christian K. "The Construction of Statistical Tables." *IRE Transactions on Engineering Writing and Speech,* EWS-5 (August, 1962), 9–14. Guidelines for constructing statistical tables, criteria for choosing vertical or horizontal presentations, and rules for the use of captions.

Aziz, A. "Article Titles as Tools of Communication." *Journal of Technical Writing and Communication,* 4 (Winter 1974), 19–21. Suggestions for forming titles that contain as much information in as few words as possible.

Bergwerk, R. J. "Effective Communication of Financial Data." *Journal of Accountancy,* 129 (February, 1970), 47–54. Suggestions on using tables to compare several pieces of data and graphs to show trends.

Bernheim, Mark. "The Written Job Search—Doubts and 'Leads.'" *The Technical Writing Teacher,* 10 (Fall 1982), 3–7. Practical strategies for job seekers.

Biskind, Elliot. "Writing Right." *New York State Bar Journal,* 42 (October, 1970), 548–554. Suggestions for improving the quality of legal writing.

Booth, Vernon. "Writing a Scientific Paper." *Biochemical Society Transactions,* 3 (1975), 2–26. A discussion of prewriting techniques, style, visual aids, and manuscript submission guidelines.

Bram, V. A. "Factors Affecting the Readability of Scientific and Engineering Texts." *The Communicator of Scientific and Technical Information,* 33 (October 1977), 3–5. Suggestions for using effective sentences as a basis for successful communication with readers regardless of their expertise.

Bromage, Mary C. "Gamesmanship in Written Communication." *Management Review,* 61 (April, 1972), 11–15. Techniques for improving business writing diction.

Christenson, Larry. "How to Write an Office Manual." *American Business,* 25 (July, 1955), 26–27, 32–34. Nine principles for writing effective manuals.

Cortelyou, Ethaline, "Abstract of the Technical Report." *Journal of Chemical Education,* 30 (October, 1955), 523–533. A discussion of the differences between abstracts and summaries.

Darian, Steven. "Using Spoken Language Features to Improve Business and Technical Writing." *ABCA Bulletin,* 44 (September, 1982), 25–30. Suggestions on the ways the differences between spoken and written communication can be used to improve business and technical writing.

Davidson, Jeffrey. "How to Spot a Phony Resume." *Supervisory Management,* 28 (May, 1983), 40–43. Guidelines for personnel managers that can serve as a warning to job applicants when preparing resumes and cover letters.

Davis, Jeffrey. "Protecting Consumers from Overdisclosure and Gobbledygook." *Virginia Law Review,* 63 (October, 1977), 841–920. A thorough discussion by a lawyer of the problems excessive disclosure, extraneous clauses, and intractable language cause consumers when they must agree to credit contracts.

DeBakey, Lois. "The Persuasive Proposal." *Journal of Technical Writing and Communication,* 6 (Winter 1976), 5–25. A discussion of the skills needed by investigators who write research proposals.

DeBeaugrande, Robert. "Information and Grammar in Technical Writing." *College Composition and Communication,* 28 (December, 1977), 325–332. A discussion of how grammar helps writers organize information.

DeRoche, W. Timothy. "Addressing Mail in the Eighties." *Balance Sheet,* 62 (February, 1981), 219–222. General recommendations for addressing mail that conform to the preferences of the U.S. Postal Service.

Deutsch, Arnold R. "Does Your Company Practice Affirmative Action in Its Communication?" *Harvard Business Review*, 54 (November–December, 1976), 16 and 186–187. Suggestions for eliminating sexism.

Dittman, Nancy A. "Job Winning Resumes." *ABCA Bulletin*, 46 (June, 1983), 16–22. A discussion loaded with practical advice and supported by examples of traditional, functional, and mixed resumes.

Dodge, W. J. "Writing the Appropriation Request." *Chemical Engineering*, 73 (March 14, 1966), 152–154. Guidelines for requesting funding.

Dolphin, Robert, Jr., and Robert A. Wagley. ".'Reading the Annual Report." *Financial Executive*, 45 (June, 1977), 20–22. Suggestions for improving the readability of annual reports.

Dover, J. C. "How to Tell Your Profits Story." *Nation's Business*, 48 (May, 1960), 38–39, 117–120. Tips for avoiding harmful confusion about the size and use of profits.

Dulek, Ron. "To Question or Not to Question: A Study of Wordiness and Impact." *ABCA Bulletin*, 45 (March, 1982), 11–14. A suggestion on using questions initially in correspondence to ensure economy in writing.

Elliott, Colin R. "Must Scientific English Be Dull?" *English Language Teaching Journal*, 31 (October, 1976), 29–34. Suggestions for producing lively scientific and technical writing.

Encke, C. G. "Scientific Writing: One Scientist's Perspective." *English Journal*, 67 (April, 1978), 40–43. Comments on the rhetorical problems that accompany the use of illustrations and figures in scientific writing.

Evered, James. "How to Write a Good Job Description." *Supervisory Management*, 26 (April, 1981), 14–19. A process approach to writing effective job descriptions and meaningful performance standards.

Fielden, John S. "What Do You Mean I Can't Write?" *Harvard Business Review*, 42 (May–June, 1964), 144–156. A discussion of readability, appropriateness, grammatical correctness, and thought content.

———— "What Do You Mean You Don't Like My Style?" *Harvard Business Review*, 60 (May–June, 1982), 128–138. Suggestions for improving effectiveness by varying style to suit individual writing situations.

Flesch, Rudolf. "How to Say It with Statistics." *Printers' Ink*, 233 (December 8, 1950), 23–24. Twelve suggestions for making statistics readable.

Francis, Henry E. "The Literary Aspects of Business Writing." *Journal of Business Communication*, 4 (October, 1966), 13–18. A discussion of form, style, and tone.

Guccione, Eugene, "Preparing Better Flowsheets." *Chemical Engineering*, 73 (March 14, 1966), 155–157. Suggestions about how to design complete, accurate, and easy-to-follow flowsheets.

Halatin, T. J. "Writing Letters of Reference That Get Results." *Supervisory Management*, 25 (December, 1980), 32–34. Seven tips for writing useful references.

Hanna, J. S. "Six Steps Toward Better Charts." *Technical Communication*, 29 (Third Quarter 1982), 4–8. Practical advice on the design, construction, and revision of tables and charts.

Happ, W. W. "Guidelines and Checklists for Preparing the Planning Report." *IEEE Transactions on Professional Communication*, PC-16 (December, 1973), 209–214. A checklist of more than 100 items.

Harty, Kevin J. "Some Guidelines for Saying 'No.' " *ABCA Bulletin*, 44 (December, 1980), 23–25. Five suggestions on how to soften the impact bad-news messages have on readers.

Hays, Robert. "Model Outlines Can Make Routine Writing Easier." *Technical Communication*, 29 (First Quarter 1982), 4–7. Models for progress or status reports, trip or site visit reports, experiment test or lab reports, investigation reports, analyses of competing ideas or plans, minutes of or confirmation reports on meetings, and long research or feasibility studies.

Jordan, Michael P. "As a Matter of Fact." *Journal of Business Communication*, 15 (Winter 1978), 3–11. A discussion of the distinctions among facts, opinions, and assumptions.

Kiritz, Norman. "Program Planning and Proposal Writing." *Grantsmanship Center News*, 7 (May–June, 1979), 33–79. Detailed, step-by-step instructions designed to assist both grant-making agencies and applicants for funding; the format suggested has been adopted by both government agencies and private foundations.

Krantz, Shelley. "Five Steps to Making Performance Appraisal Writing Easier." *Supervisory Management,* 28 (December, 1983), 7–10. Guidelines for writing job appraisals that will have a positive effect on employee performance.

Limaye, Mohan R. "Approaching Punctuation as a System." *ABCA Bulletin,* 46 (March, 1983), 28–33. An argument that punctuation is a rational system of easy-to-learn rules rather than a conglomeration of seemingly unrelated rules and exemptions to rules.

Messer, Donald K. "Six Common Causes of Ambiguity." *The Technical Writing Teacher,* 7 (Winter 1980), 50–52. Implication, word order, ambiguous words, dangling participles, missing or improper punctuation, and faulty reference.

Motley, Robert J. " 'Sexism' and Modern Business Communications." *Business Education Forum,* 33 (November, 1978), 28–31. Eight suggestions for eliminating sexism in business communication.

Palmer, Stacy E. "What to Say in a Letter of Recommendation." *Chronicle of Higher Education,* 27 (September 7, 1983), 21–22. Guidelines to ensure that recipients of written recommendations receive the information they really need about applicants.

Peterson, Becky K. "Tables and Graphs Improve Reader Performance and Reader Reaction." *Journal of Business Communication,* 20 (Spring 1983), 47–55. Results of a study suggesting how effective tables and graphs improve reader comprehension of written materials.

Plung, Daniel L. "Writing the Persuasive Business Letter." *Journal of Business Communication,* 17 (Spring 1980), 46–49. Guidelines for arranging ideas in a pattern corresponding to a reader's decision making processes.

Richards, Paul. "Sentence Control: Solving an Old Problem." *Supervisory Management,* 25 (May 1980), 37–43. Suggestions for improving readability by varying sentence length and punctuation.

Samora, Julian, and others. "Medical Vocabulary Knowledge among Hospital Patients." *Journal of Health and Social Behavior,* 2 (1961), 83–92. Results of a study with important implications about the problems lay audiences can have with even basic specialized terminology.

Schindler, George E., Jr. "Why Engineers and Scientists Write as They Do—Twelve Characteristics of Their Prose." *IEEE Transactions on Professional Communication,* PC–18 (March, 1975), 5–10. A discussion of structural and grammatical problems commonly found in the sentences engineers and scientists write.

Selzer, Jack. "Readability Is a Four-Letter Word." *Journal of Business Communication,* 18 (Fall 1981), 22–34. An argument against the validity of readability formulas.

Shelby, Anne. "How to Type Your Paper." *The Technical Writing Teacher,* 1 (Winter 1974), 11–22. Easy-to-follow directions for typing formal reports.

Spangler, E. R. "Modern Grammar and Its Application to Technical Writing." *Journal of Chemical Education,* 33 (February, 1956), 61–64. An application of some basic principles of descriptive grammar to technical writing.

Sparrow, W. Keats. "Six Myths about 'Writing for Business and Industry.' " *The Technical Writing Teacher,* 3 (Winter 1976), 49–59. Practical advice for beginning teachers of business and technical writing.

Stiegler, Christine B. "Are You a Closet Sexist?" *Supervisory Management,* 24 (May 1979), 38–41. A brief discussion of sexism in verbal and nonverbal communication.

Stine, Donna. "The Writing of Performance Appraisals: A Survey of Personnel Directors." *ABCA Bulletin,* 45 (March, 1982), 29–33. A discussion of common weaknesses in, goals for, and potential legal problems with written performance appraisals of employees.

Storm, A. Allman. "In Search of Money." *Management Accounting,* 53 (August, 1971), 29–32. Guidelines for those seeking capital.

Swift, Marvin. "Writing a Problem-Centered Report." *ABCA Bulletin,* 46 (September, 1983), 19–23. Guidelines for investigating and reporting on problems.

Tibbetts, Arn. "Ten Rules for Writing *Readably.*" *Journal of Business Communication,* 18 (Fall 1981), 53–62. Suggestions on how to write like "a human being."

Van DeWeghe, Richard. "Writing Models, Versatile Writers." *Journal of Business Communication,* 20 (Winter 1983), 13–23. Five models of the composing process.

Waters, Max L. "Abstracting—An Overlooked Management Writing Skill." *ABCA Bulletin,* 45 (June, 1982), 19–21. A general discussion distinguishing among descriptive, informative, analytical, and critical abstracts.

Weightman, Frank C. "The Executive Summary: An Indispensable Management Tool." *ABCA Bulletin,* 45 (December, 1982), 3–4. Suggestions for using effectively written executive summaries to involve the reader immediately in business reports.

A 4
B 5
C 6
D 7
E 8
F 9
G 0
H 1
I 2
J 3